990

Research Agenda for Mathematics Education

Research Issues in the Learning and Teaching of Algebra

Volume 4

Editors

Sigrid Wagner
University of Georgia

Carolyn Kieran
Université du Québec à Montréal

LAWRENCE ERLBAUM ASSOCIATES

**NATIONAL COUNCIL OF
TEACHERS OF MATHEMATICS**

ISBN: 0-87353-268-6 (Vol. 4, paper)
ISBN: 0-8058-0354-8 (Vol. 4, cloth)
ISBN: 0-87353-256-2 (5-vol. set, paper)

Second printing 1989

This material is based upon work supported by the National Science Foundation under Grant No. MDR-8550614. The Government has certain rights in this material. Any opinions, findings, and conclusions or recommendations expressed in this material are those of the author(s) and do not necessarily reflect the views of the National Science Foundation.

CONTENTS

Acknowledgments

Neither the conference nor the monograph would have been possible without the generous support of many individuals. We first want to thank conference participants who took several days from their busy schedules to come to Georgia. Their names are listed at the back of this volume. Extra thanks go to those participants who spent countless hours writing papers for the conference and subsequently revising them for this monograph.

We also wish to thank the following participants and observers for their careful note-taking during the conference: John Bernard, Charlee Goolsby, Larry Hatfield, Mary Lindquist, Francis Norman, George Springer, Jane Swafford, and Beverly Whittington. Our thanks are extended to the following graduate students who helped with the day-to-day running of the conference: Adalira Ludlow, Jose Matos, Robert Moore, Julio Mosquera, Tzyh-Chiang Ning, Yasuhiro Sekiguchi, Siriporn Thipkong, Catherine Vistro, and Joseph Zilliox. We are grateful for the secretarial assistance provided by Stephanie Bales and Valerie Kilpatrick during the conference. And, we very much appreciate the administrative and financial support provided for the conference by Dr. Alphonse Buccino, Dean of the College of Education, and Dr. James W. Wilson, Head of the Department of Mathematics Education, University of Georgia.

A large measure of thanks goes to Lise Monaghan, Carolyn Pittenger, and Suzanne Turcotte for their editorial assistance in the preparation of the monograph, and to our graphics artist Lorraine Lavigne. We acknowledge the administrative and financial support provided by Dr. Robert Anderson, Chairman of the Département de Mathématiques et d'Informatique, Université du Québec à Montréal. Finally, the entire project would not have been possible without the financial support of the National Science Foundation and the continued assistance of the National Council of Teachers of Mathematics. Our thanks to all of you.

Sigrid Wagner
Carolyn Kieran

Series Foreword

We clearly know more today about teaching and learning mathematics than we did twenty years ago, and we are beginning to see the effects of this new knowledge at the classroom level. This is possible in part because of the financial support that has become available to researchers. If theory building and knowledge acquisition are to have a basis broad enough to inform policy and influence educational practice, such support is essential. Although funding levels remain low in comparison to existing needs, there are several research projects either completed or in progress that could not have been undertaken without support.

In particular, we can point to several significant sets of studies based on emerging theoretical frameworks. For example, young children's early number learning and older children's understanding of rational numbers have been the subject of several recent research programs. Most of us who do research would agree that our work is more likely to be profitable when it results from an accumulation of knowledge acquired through projects undertaken within a coherent framework rather than through single, isolated studies. To establish such a framework, researchers must be provided with the opportunity to exchange and refine their ideas and viewpoints. Conferences held in Georgia and Wisconsin during the seventies serve as examples of the role such meetings can play in providing a vehicle for increased communication, synthesis, summary, and cross-disciplinary fertilization among researchers working within a specialized area of mathematical learning.

Over the past few years, the members of the Research Advisory Committee of the National Council of Teachers of Mathematics (NCTM) have observed specializations emerge that could benefit from collaborative efforts. We therefore proposed to the National Science Foundation that funding be provided for the purpose of establishing research agendas in several areas where conceptual and methodological consensus seemed possible. We believed that such a project was needed at this time for two reasons: first, to direct research efforts toward important questions, and second, to encourage the development of support mechanisms essential to collaborative chains of inquiry. Four such specialized areas were selected for this project: the teaching and assessing of problem solving, the teaching and learning of algebra, effective mathematics teaching, and the learning of number concepts by children in the middle grades.

The plan for the project included a working group conference in each of the four areas, with monographs of conference proceedings to be published by the National Council of Teachers of Mathematics. An overview monograph, written by advisory board members, was also planned. The advisory board consisted of F. Joe Crosswhite, James G. Greeno, Jeremy Kilpatrick, Douglas B. McLeod, Thomas A. Romberg, George Springer, James W. Stigler, and Jane O. Swafford, while I served as project director. For each of the four selected areas, two researchers were named to serve as confer-ence co-directors and as co-editors of the monograph of conference proceedings. These pairs were Edward A. Silver and Randall I. Charles for teaching and assessing problem solving; Sigrid Wagner and Carolyn Kieran for learning and teaching algebra; Douglas A. Grouws and Thomas J. Coo-ney for effective mathematics teaching; and Merlyn J. Behr and James Hiebert for number learning in the middle grades.

The project began in May of 1986 with a planning meeting of advisory board members and conference co-directors. Issues to be addressed and possible paper topics for each conference were first identified by the group. Tentative lists of invitees for each conference were drawn up to include researchers from mathematics education and relevant fields of psychology and social science, as well as mathematicians and practitioners. The names of promising young researchers were included along with names of estab-lished researchers. An international perspective was considered important, and so the list also included names of scholars from abroad. The concept of working group conferences funded for 25 people precluded expanding the conference to all interested persons. We therefore decided to invite people to attend only one conference, thus maximizing the total number of persons involved in the project. The final participant lists follow this report.

The first of the four working conferences, on teaching and assessing mathematical problem solving, was held in January, 1987, in San Diego. Several approaches to teaching problem solving were advanced and dis-cussed: teaching *as* problem solving, the teacher as coach versus teacher as manager, modeling master teachers, viewing students as apprentices, the use of macro-contexts to facilitate mathematical thinking, consideration of mathematics as an ill-structured discipline. Discussion of assessment issues focused on process rather than outcome, and questions of evaluating pro-cessess and schema structures were explored. Considerable attention was given throughout the conference to associated problems of teacher prepa-ration.

The conference on effective mathematics teaching was held in March in Columbia, Missouri. The first paper delivered there, on teaching for higher-order thinking in mathematics, tied this conference to the first one. A paper on the functioning of educational paradigms set a stage for much of the discussion at the conference. The question of what makes a good mathe-matics teacher was explored in discussions ensuing from presentations on expert-novice studies, on cross-cultural studies, and on teacher profession-alism. Concern for the content in mathematics classes and the manner in which this is determined also received attention. The lack of funding nec-essary for observational research was strongly noted.

The conference on teaching and learning algebra, also in March, was held in Athens, Georgia. The papers and discussions focused on four major themes: what is algebra and what should it become, in light of continuing technological advances; what has research told us about the teaching and learning of algebra; what is algebraic thinking and how does it relate to general mathematical thinking; and what is the role of representations in the learning of algebra. Research questions from the perspectives of content, of learning, of instruction, and of representations were formulated.

The final conference, on number concepts in the middle grades, was held in May in DeKalb, Illinois. A theme permeating many of the papers at this conference was that number concepts related to topics taught at this level are qualitatively different from those in lower grades, both in terms of the number systems studied and the operations involved. The papers explored the new conceptions and complexities that students encountered, and examined the effects of conventional and experimental instructional programs. It was acknowledged that the differences between early and later number concepts need to be recognized and more adequately understood before instructional programs can be developed to enhance number learning in the middle grades.

The brevity of these descriptions does not do justice to the diversity and richness of the papers and discussions. Each of the conferences was indeed a working conference. Participants addressed difficult questions and discussions were lively and intense. Consensus was elusive, as might be expected with a group of people with such diverse backgrounds. Even so, there was agreement on many fundamental issues, and individual researchers, representing different disciplines and viewpoints, were able to reach new understandings.

There are five monographs being published as a result of this project, all under the general title *Research Agenda in Mathematics Education.* Four of the monographs contain conference proceedings and are subtitled as follows: *The Teaching and Assessing of Mathematical Problem Solving, Perspectives on Research on Effective Mathematics Teaching, Research Issues in the Learning and Teaching of Algebra,* and *Number Concepts and Operations in the Middle Grades.* The proceedings include revised conference papers, some discussion, response, and summary papers by other conference participants, and chapters by the co-editors. They will be of particular interest to researchers interested in the learning and teaching of mathematics. The fifth monograph, *Setting a Research Agenda,* is intended for a wider audience, including policy makers, mathematics supervisors, and teachers. In this monograph the project advisory board discusses the past and present state of research in mathematics education and cognitive science, the relation of reform movements to research efforts, the role of this project in guiding future research in mathematics education, major issues addressed at the conferences and other issues still needing to be addressed, and resources needed to facilitate research. We are fortunate to have Lawrence Erlbaum Associates, Inc., join with the National Council of Teachers of Mathematics in publishing this series of monographs.

This project was funded by the National Science Foundation under Grant No. MDR-8550614. The advisory board and monograph editors join me in an expression of gratitude to Raymond I. Hannapel of the National Science Foundation for his continued support throughout the term of this project. We also wish to thank James D. Gates and Charles R. Hucka of the National Council of Teachers of Mathematics for their assistance with the publication of these monographs, and Julia Hough of Lawrence Erlbaum Associates for her work in facilitating joint publication between NCTM and Erlbaum. Finally, we wish to acknowledge the assistance of administrators at San Diego State University, University of Missouri, University of Georgia, and Northern Illinois University, and thank them for the many amenities they provided to conference participants.

I personally want to express my appreciation to the members of the advisory board for all the assistance they have given me during these past two years. My largest debt of gratitude is owed to the conference co-directors and co-editors for their work in directing four outstanding conferences and for providing all of us with sets of proceedings that will guide research efforts in the years to come.

Judith T. Sowder
San Diego State University

The Research Agenda Conference on Algebra: Background and Issues

Carolyn Kieran
Département de Mathématiques et d'Informatique
Université du Québec à Montréal

Sigrid Wagner
Department of Mathematics Education
University of Georgia

THE TEACHING AND LEARNING OF ALGEBRA DURING THE PAST CENTURY

In the 1880s, when only one youth in ten attended high school, algebra was a required course of study in the first year. The topics covered generally included operations with literal expressions, the solving of both linear and quadratic equations, the use of these techniques to find answers to problems, and practice with ratios, proportions, powers, and roots. However, the algebra curriculum during this period varied from one school or state to another because each college and university had its own individual entrance examination and thereby dictated the school syllabus to be followed by prospective students. The situation changed in 1900 in the United States when Columbia University took steps to set up the College Entrance Examination Board (CEEB). The advisory council of this board, which included mathematicians and college preparatory teachers, began work on syllabi that would be the basis for uniform, nationwide algebra examinations. Textbooks were closely patterned after the CEEB syllabi.

The situation in other countries was not quite the same, however. In England, for example, John Perry, a professor of mechanics and mathematics at the Royal College of Science in London, in 1901 formed a committee of mathematicians, school principals, and teachers to investigate means of improving the teaching of mathematics. The report of this committee formed the basis for a new syllabus proposed by Perry. He advocated much laboratory work and manipulation of concrete materials by combining the teaching of mathematics and science, with emphasis on the practical side of mathematics, that is, arithmetic computations, mechanical drawing, and graphical methods.

Thus, pure mathematicians had a stronger influence on the high school algebra curriculum in North America in the first decade of the 20th century than they had in England. Yet, even during this period, there were undercurrents of protest in the United States. By the early twenties, the forces of change were too great to be ignored, forces such as a surge in school

1

enrollment and cries for a more practical, vocationally oriented curriculum. In many states, child-labor laws and compulsory school attendance laws kept students in school for at least ten years. Thus, farm youths who had hitherto stayed away from any form of secondary education were now taking algebra. The proportion of students attending high school grew from 1 in 10 to 1 in 3.

This change in the school population led to changes in emphasis in the algebra curriculum. The traditional, formal approach—in which algebra was taught as a purely mathematical discipline, with no emphasis on understanding—survived a while longer in the CEEB examinations but was considered a thing of the past among leaders in the teaching of mathematics. The newer, more practical approach to algebra that had taken hold in England was beginning to appear in many of the algebra textbooks in the United States too. Eventually, the College Entrance Examination Board called a group of educators together to consider modifying its algebra syllabi. Yet the hold of the CEEB on the formal algebra curriculum was not broken until 1930 when the Great Depression precipitated renewed attacks on education in general, and mathematics in particular, for being so far removed from everyday life.

However, the new approach to algebra was not without its critics. The educational psychologist, Edward L. Thorndike, argued in *The Psychology of Algebra* (Thorndike, Cobb, Orleans, Symonds, Wald, & Woodyard, 1923) that, even though he preferred the new algebra curriculum, some of its features could cause confusion in the minds of learners. For example, in the algebra of a generation earlier, an equation was a thing to be solved and nothing more. With the newer approach, an equation was also considered an expression of a general relation among variables, involving at times a Cartesian graph. Thorndike argued that these two aspects of equations should be kept distinct at the start and, to a large extent, throughout algebra. His prescriptions for teaching algebra included consideration of which "bonds" should be formed, the amount of practice students should have, and how this practice should be distributed. Although Thorndike's analysis of algebra in terms of bonds and the accompanying prescription of repetitive practice to promote learning fell into disfavor, he must be credited with bringing to algebra research a scientific approach that involved careful analysis of the nature of the task.

What we know of the learning of algebra from research during the years 1900 to 1930 deals primarily with the relative difficulty of solving various kinds of linear equations. Researchers of this period (e.g., Hotz, 1918; Monroe, 1915; Reeve, 1926) tended to give their subjects timed equation-solving tests. They reported the number of attempts and the number of correct solutions. All that seems clear from these studies is that the larger the number of steps required to solve an equation, the less often students were able to solve it correctly.

The primacy of Thorndike's connectionist learning theory was challenged in the mid-1930s by meaningful learning theory, as set forth by Brownell (1935). Even though Thorndike had indicated the importance of understanding algebraic formulas before doing manipulation exercises, he had clearly put the emphasis on practice. He had also proposed that practice often led to understanding. These ideas were disputed by the meaningful learning theorists and educators of the 1930s and later, who maintained that understanding could not be achieved by drill alone and that understanding should be sought prior to practice.

It is also important to note that meaningful learning theorists were generally more interested in the *processes* of learning than in the *products* of learning. Unfortunately, as Weaver and Suydam (1972) have pointed out, meaningful learning theory was never clearly interpreted at the secondary school level, only at the level of elementary school arithmetic. Thus, very little research related to the processes involved in the learning of algebra was carried out. In one study that solicited verbal responses from high school students, Breslich (1939) analyzed the errors made in applying some of the algorithms of first-year algebra. However, most of the procedures investigated had become so routinized that the researcher was not able to identify the processes involved in the students' learning.

Despite the contributions of meaningful learning theorists to the teaching of school mathematics during the period from 1930 to 1950, the main influence on the teaching of algebra during this time was the socio-economic conditions produced by the Great Depression and World War II. Preoccupation with issues of existence and survival led to the entrenchment of a kind of mathematics that was practical and vocationally oriented. The progressive education movement and the growing interest of psychologists in the study and measurement of individual differences contributed to an increasingly child-centered focus in the classroom. Moreover, compulsory school attendance had become universal, so high schools were faced with educating everyone. They responded by lessening graduation requirements and developing more elective courses. Enrollments in the traditional mathematics courses declined precipitously, and curriculum innovators developed courses like consumer mathematics to appeal to the majority of the population. University mathematicians, of course, had little interest in such developments. In fact, the period from 1930 to 1950 was the only period in the last hundred years when none of the important North American mathematics education committees included leading research mathematicians (Rosskopf, 1970).

The child-centered focus in mathematics classrooms began to shift in the 1950s. World War II had demonstrated how inadequate the mathematics preparation of our young people was. After the war, the requirements of industry in engineering, the sciences, and mathematics escalated dramatically. The gap between the mathematics taught in high schools and the mathematics required for postcollege jobs in such fields as nuclear physics, space exploration, communications, and computers grew wider year by year. Cold War fears fueled concerns about our preeminence in mathematics, science, and technology. The appointment in 1955 of the CEEB Commission on Mathematics, the launching in 1957 of the Soviet Sputnik, and the founding in 1958 of the School Mathematics Study Group marked the beginning of an era in which the voices of mathematicians would, once again, be heard.

In addition to major technological advances, significant developments had been occurring in mathematics itself over the past century or so. Thus it was not difficult to find support for the idea that the mathematics taught in school should reflect the new ideas that mathematicians were finding so useful and powerful. The emphasis in a revised school mathematics program was to be on deductive reasoning, inductive discovery, and precision of language. The main objective was the development of ideas and understanding rather than merely computational skill and the recall of seemingly unrelated facts. Algebra in the new curriculum was to incorporate new topics like inequalities, emphasize unifying concepts like set and function, and be taught so that its structure and deductive character were apparent.

But teachers had to be retrained to teach this "new math." Summer institutes attracted top-notch mathematics teachers who were interested in learning more about the structural approach to mathematics and how they might present it to their students. Some of these teachers continued on to earn master's degrees and doctorates in new programs that focused specifically on mathematics education.

With the explosion in graduate education came a dramatic increase in research activity. During the fifties and early sixties, research related to algebra had been conducted mostly by psychologists with a behaviorist orientation, who used mathematics as a convenient vehicle for studying general questions related to concept attainment, skill development, memory, and the like. By the late sixties, research in mathematics education was increasingly being conducted by mathematics educators, who had strong backgrounds in the mathematical content and keen interest in the psychological problems of learning and teaching mathematics. That mathematics education was beginning to be recognized as a distinct professional field was reflected in the founding of two journals devoted to mathematics education research (*Educational Studies in Mathematics* in 1968 and the *Journal for Research in Mathematics Education* in 1970) and the holding of the first International Congress on Mathematical Education in 1969.

Much of the mathematics education research of this period was Piagetian in approach. Many of Piaget's writings had been translated into English only in the 1950s, and mathematics education researchers, as experienced classroom teachers familiar with the day-to-day challenges of communicating mathematical ideas, found much that was appealing and/or intriguing in Piaget's theory. The influence of cognitive development psychology on early mathematics education research was clearly evident. Several studies attempted to apply, for example, Piaget's notion of stages to the learning of basic mathematical concepts, including some algebraic concepts like function and variable. The insights from these psychologically oriented studies had, however, little immediate impact on the school mathematics curriculum—they were no match for the bulldozing effect of the 1970s back-to-basics backlash against the new math movement.

As we find ourselves moving now into a new era, the mathematics curriculum today reflects some lasting changes wrought by the mathematicians of the sixties, major impact from the socio-political back-to-basics movement, and significant influence from mathematics educators in the current emphasis on problem solving, estimation, measurement, and other "new" basic skills. However, many curriculum innovators seem to be listening to other voices too. One of these other voices is that of the computer technologist. A glance at some of the newest algebra textbooks that feature computer tools and methods shows the potential impact of technology on the teaching of algebra.

Another voice that curriculum developers seem willing to listen to at the current time is that of the psychologist. Because a curriculum based on sound mathematical notions and sequencing did not produce the desired effects for many students in the sixties, curriculum developers are looking increasingly to research to provide insights into the nature of the learning of school mathematics. The rise of information processing psychology has led to an emphasis by many psychologists on the study of the cognitive processes underlying the learning of complex tasks, especially those in the area of school mathematics. At the same time, the maturing of mathematics education as an interdisciplinary field of study has led many mathematics education researchers to begin melding a strong content orientation to a sharpening focus on cognitive processes. Thus, we see a growing research interest in the factors and processes involved in the learning of school algebra merging from two directions, from cognitive psychologists and from mathematics educators.

This has been an all-too-brief summary of some of the main trends in algebra teaching, learning, and research over the past hundred years. But it was this history that formed the backdrop for the Research Agenda Conference on the Learning and Teaching of Algebra.

THE RESEARCH AGENDA CONFERENCE ON ALGEBRA

In March 1987 a four-day working conference on research in the learning and teaching of school algebra was held at the University of Georgia. This conference, which was part of the Research Agenda Project (see Series Foreword, this volume) conducted by the National Council of Teachers of Mathematics, brought together mathematicians, mathematics educators, psychologists, technologists, researchers, practitioners, and curriculum developers—all with special interest and background in the area of algebra. The conference co-organizers invited participants from several disciplines, in order to provide for different perspectives on research and its implications for the learning and teaching of algebra.

Some of the general questions and issues underlying the structure of the planned program were: (a) What *is* algebra? (b) What *should* algebra be, particularly in view of continuing technological advances? (c) What does research say about the learning and teaching of algebra? (d) What are the implications for the classroom and for curriculum development? and (e) Which learning/teaching theories guide our research?

Certain participants were invited to present papers related to these issues, and other participants were asked to react to these papers. Interspersed among the presentations were both large- and small-group discussions. The large-group discussions, led by a discussant, generally focused on the presentation that had just taken place and provided a forum for participants to state how the ideas presented informed their own research. The small-group discussions provided an opportunity to identify commonalities among the various presentations and to analyze and synthesize ideas that had been put forward.

The conference opened with a mathematician's perspective on the nature of algebra and on some of the factors involved in the learning of algebra at the school level. John Thorpe examined several topics for possible inclusion in a modified algebra curriculum using the three criteria of intrinsic value, pedagogical value, and intrinsic excitement. In order to provide the group with another mathematician's perspective, a reaction to the opening session was given by a second mathematician, Joan Leitzel. She emphasized the need to change our approach to teaching algebra, stressing that the availability of calculators and computers permits a numerical approach to algebra that is quite different from the formal axiomatic approach of recent years.

These two presentations were followed by two reviews of the research literature by mathematics education researchers, one covering the early learning of algebra, the other the later learning of algebra. Carolyn Kieran reviewed studies that have focused on students' learning of variables, algebraic expressions, equations, and equation solving—with particular emphasis on the difficulties learners encounter with some of the structural features of school algebra, tracing some of those difficulties back to experiences in arithmetic. Lesley Booth, in her reaction to the Kieran paper, emphasized the distinction that ought to be made between the syntactic and semantic aspects of algebra.

Nicolas Herscovics reviewed the research literature related to the learning of equations in two variables, their graphs, and the notion of function. These topics were viewed through the lens of the cognitive obstacles that are involved in learning algebra. David Tall's reaction to this presentation focused primarily on the relevance of pre-computer research findings to a new computer paradigm.

The conference program continued with two presentations from cognitive psychologists. Seth Chaiklin reviewed the cognitive psychology literature on algebra learning, with particular emphasis on research related to the solving of algebra word problems. He pointed out that the perspective of psychologists who study algebra problem solving is generally to proceed in terms of what can be done, as opposed to what is needed; thus, he suggested that some sensitivity is required on the part of mathematics educators who are interested in using these research results to guide their practice. The reaction, provided by Bob Davis, underlined the artificiality of typical algebra word problems and suggested a more fundamental kind of research if we are to make progress in changing the content of algebra courses.

The second presentation from a cognitive science perspective was provided by Jill Larkin. In a paper that addressed the symbol system of algebra, she described the development of a model of competent performance in algebra and proposed testing the model by using it to design instruction. Mary Grace Kantowski, in her reaction to the Larkin paper, pointed out some of the weaknesses of a rule-driven interpretation of algebra learning.

The potential role(s) and impact of technology were taken up in subsequent presentations. Pat Thompson considered recent developments in intelligent tutoring systems (ITSs) and reviewed several current projects that illustrate the power of concepts and methodology from artificial intelligence (AI) for developing systems that present algebraic content in substantially new ways. However, in cautiously describing the power of AI for task and concept analysis, he also pointed out that a rule-based conception of competence in algebra fails to look at incompetence as stemming from impoverished conceptual knowledge. In his reaction to the Thompson paper, Matt Lewis addressed several of Thompson's objections to ITS research and clarified the purpose and value of rule-based simulations as a first step in building theories of human learning.

Jim Kaput's presentation focused on the issue of representation in algebra, in particular, the use of technology to support simultaneous, multiple representations of algebraic subject matter. He described some inroads that have been made in developing software that incorporates multiple, linked representations but also pointed out that algebra learning research has not been able to keep step with technology. David Kirshner's reaction to the Kaput paper questioned whether Kaput's theoretical framework, which is clearly of value for analyzing and categorizing software environments, can also serve more broadly as a metalanguage for research into the psychology of algebra.

Consideration of the issues related to curriculum development began with a roundtable session involving three mathematics educators from different parts of the world, who described aspects of the algebra programs with which they come into contact on a day-to-day basis: Tatsuro Miwa (Japan), Eugenio Filloy (Mexico), and Sid Rachlin (Hawaii).

The focus on curriculum development continued with a presentation by Jim Fey on the impact that recent and future developments in computer technology should have on our conception of algebra. He suggested dramatic ways of modifying the curriculum, based on an assumption of universal access to the new technology, and proposed that we now have the tools that enable us to modify our skill-dominated conception of school algebra and rebalance it in favor of objectives related to understanding and problem solving. In her reaction to Fey's proposals, Sharon Senk suggested that the reforms Fey described will have limited impact on algebra by the year 2000 unless the mathematics education community deals with the four factors of teachers, textbooks, evaluation, and articulation between high school and college coursework.

The above paper presentations, reactions, and discussion sessions constituted the first part of the conference program. The second part of the program dealt explicitly with the generation of issues to be considered for a research agenda in algebra—a process that had begun implicitly from the time the conference started. This proved to be a rather complex task, particularly because all of us were attempting to project ourselves into a future in which changes will occur with increasing rapidity. The uncertain relationship between school algebra and technology made it difficult to design a multi-faceted research program that would retain its relevance throughout an extended period of time. Nevertheless, many of the research issues that were generated reflected conference participants' concern with the impact of technology on school algebra. Other issues covered a wide range of concerns related to content, learning, teaching, algebraic thinking, affect, representation, curriculum development, testing, and teacher education.

Despite the fact that the agenda was far from being in its final form by the last day of the conference, draft copies of the agenda were circulated to participants, and the ideas that had been generated served as a basis for two closing panel sessions, the third and last part of the program. One panel provided four different perspectives on the emerging research agenda: (a) that of a mathematics educator, Lesley Booth; (b) that of a cognitive scientist, Matt Lewis; (c) that of a practitioner, Diane Briars; and (d) that of a curriculum developer, Sid Rachlin. Some of the main issues that these panelists tended to emphasize were implementation issues. For example, it was pointed out that a primary goal of mathematics education research has always been to have an impact on classroom practice. If research is to have an impact, teachers must be brought into the research process. It was also suggested that research issues and results be incorporated into preservice and in-service teacher education courses. But, as was pointed out, more than research studies and results are needed. It is necessary to have curriculum and assessment materials that reflect the implications of research.

Other issues raised by the panelists concerned the kinds of algebraic skills and understandings required by today's students, and what teachers can and should do with technology.

The other closing panel session focused on the issue of theoretical and conceptual frameworks in research on the teaching and learning of algebra. Three panelists from different research traditions—Bob Davis, Jill Larkin, and David Wheeler—discussed the existence/nonexistence of theory in algebra research. Some of the ideas that emerged from this exchange concerned the two-way relationship between theory and practice and the need to build theories based on the experience of practitioners. It was also suggested that new theories need to be constructed in order to attempt to tie together the results of researchers from different traditions and to be able to predict how algebra learning takes place.

All too soon, the conference came to an end. Those who had presented papers and reactions were invited to integrate into their final drafts any pertinent comments, suggestions, or ideas that had surfaced during the conference discussions. Sometime after the conference, members of the closing panel on the research agenda received a more polished version of the agenda to which they prepared their written commentaries. This monograph thus represents the fruits of the participants' pre- and post-conference reflections.

THE STRUCTURE OF THE MONOGRAPH

The monograph is structured to correspond, in general, with the sessions of the conference; thus, it is divided into three parts. Part I contains the papers and reactions related to the thematic questions of the conference, such as: (a) What is algebra? (b) What should algebra be? and (c) What does research say about the learning and teaching of algebra? These papers are sequenced in the order in which they were presented at the conference. Part II begins with an elaborated version of the research agenda that was generated by conference participants. Reactions to the agenda from three different perspectives are provided in the papers that follow. Part III concludes the monograph with three retrospective papers on theoretical issues related to algebra research.

REFERENCES

Breslich, E. R. (1939). Algebra, a system of abstract processes. In C. H. Judd (Ed.), *Education as cultivation of the higher mental processes*. New York: Macmillan.

Brownell, W. A. (1935). Psychological considerations in the learning and teaching of arithmetic. In W. D. Reeve (Ed.), *The teaching of arithmetic* (Tenth Yearbook of the National Council of Teachers of Mathematics). New York: Teachers College Bureau of Publication.

Hotz, H. G. (1918). First-year algebra scales. *Contributions to Education* (No. 90). New York: Columbia University, Teachers College.

Monroe, W. S. (1915). A test of the attainment of first-year high-school students in algebra. *School Review, 23,* 159-171.

Reeve, W. D. (1926). *A diagnostic study of the teaching problems in high school mathematics.* Boston: Ginn.

Rosskopf, M. R. (1970). Mathematics education: Historical perspectives. In M. R. Rosskopf (Ed.), *The teaching of secondary school mathematics* (Thirty-third Yearbook). Washington, DC: National Council of Teachers of Mathematics.

Thorndike, E. L., Cobb, M. V., Orleans, J. S., Symonds, P. M., Wald, E., & Woodyard, E. (1923). *The psychology of algebra.* New York: Macmillan.

Weaver, J. F., & Suydam, M. N. (1972). *Meaningful instruction in mathematics education.* Columbus, OH: ERIC Information Analysis Center for Science, Mathematics, and Environmental Education.

Part I

Past Research and Current Issues

Algebra: What Should We Teach and How Should We Teach It?

John A. Thorpe
Department of Mathematics
State University of New York at Stony Brook
and
National Science Foundation

The teaching of algebra in the schools is not significantly different today from what it was fifty years ago. Certainly, there have been *some* changes during those years. The "new math" movement of the 1960s attempted (and, briefly, succeeded) in introducing some new ideas and new approaches into algebra instruction. But the changes that have persisted into today's curriculum have been more cosmetic than substantial.

Meanwhile, mathematics and its applications have changed dramatically in the past fifty, even in the past ten, years. The emergence of sophisticated calculators and computers as tools in both computational and abstract mathematics have, in particular, changed the way that mathematicians do mathematics and the way that scientists, engineers, and social scientists use mathematics. It is time for algebra instruction in the schools to begin to reflect these changes. It is time for a major reevaluation of the content of the algebra curriculum and of the instructional strategies that are used in teaching algebra.

GOALS FOR ALGEBRA INSTRUCTION

Any such reevaluation must necessarily begin with an examination of the goals for algebra instruction. A number of goals have been proposed:

- To "develop student skills in the solution of equations, finding numbers that meet specified conditions" (Fey, 1984, p. 14);
- To teach students to use symbols to help solve real problems, such as mixture problems, rate problems, and so forth (A. Schoenfeld, remarks at the Mathematical Sciences Education Board conference,

11

"The School Mathematics Curriculum: Raising National Expectations," UCLA, November 1986);

- To prepare students to follow derivations in other subjects, for example, in physics and engineering (H. Flanders, remarks at the annual
meeting of the Mathematical Association of America, San Antonio,
January 1987)[1];

- To enable students to become sufficiently at ease with algebraic formulas that they can read popular scientific literature intelligently
(Mathematics Proficiency Committee, SUNY at Stony Brook, discussion, c.1975).

But, as Schoenfeld has pointed out, algebra should not be taught as a
collection of tricks, as is common in textbooks—a trick for this, a trick for
that. Students should see algebra as an aid for thinking rather than a bag
of tricks. Whitney (1985) carries the thinking theme one step further: Students should grow in their natural powers of seeing the mathematical
elements in a situation, reasoning with these elements to come to relevant
conclusions, and carrying out the process with confidence and responsibility.

I am in complete agreement with these goals, especially the goal articu-
lated by Whitney. All mathematics instruction, and algebra instruction in
particular, should be designed to promote understanding of concepts and to
encourage thinking. Drill and practice should be required whenever necessary to reinforce and automatize essential skills. But, whenever drill and
practice are required, students should always have a clear understanding of
why the particular skill is so important that its mastery is required.

The correctness of the goal of a "thinking curriculum" is, in my opinion,
self-evident. This general goal, however, must be supplemented by more
specific goals that provide guidance in the selection of topics to be taught.
In my view there are three criteria, at least one of which must be met before
any given topic merits inclusion in the curriculum:

- intrinsic value,
- pedagogical value, and
- intrinsic excitement or beauty.

Intrinsic Value

Some topics must be included in the curriculum because they are, or will
be, important in the lives of the students. The distance formula, percents,
graphs, and probability and statistics are examples of topics with clear
intrinsic value.

As a more detailed example, consider the concept of function. Functions
are at the very heart of calculus, and that is sufficient reason to justify the
inclusion of functions in any algebra course for the college bound. But
functions should be taught to *all* students, because the concept of function

is one of the most important of mathematical concepts. An understanding of the function that translates interest rates into monthly payments on loans is of intrinsic value to most people several times in life. The function that assigns sales tax to purchases is encountered by most people on a daily basis.

But one caveat: Let us not define a function as a set of ordered pairs! The definition of a function as a set of ordered pairs is not only too abstract for an initial introduction, it is inconsistent with the way functions are viewed and used by professionals. A function is a rule—a rule that assigns to each member of some set a member of some other (or possibly the same) set. A function does have various representations—as a set of ordered pairs, as a graph, or as a point in a function space(!)—but we should teach the most intuitive and practical definition and not confuse our students with unnecessary abstractions.

As I was preparing this paper, I reviewed three of the currently most popular algebra texts, which I shall refer to as AW (Keedy, Bittinger, Smith, & Orfan, 1984), HBJ (Coxford & Payne, 1987), and HM (Dolciani, Wooton, & Beckenbach, 1983). One of my biggest disappointments was in observing that, in two of them (HBJ and HM), the ordered pair definition of function still persists. I recognize that this approach was popular among professional mathematicians 25 years ago, and that many of them advocated using this approach in the schools, but this was certainly one of the errors of the sixties and it is time that it were laid to rest.

At the risk of belaboring the point, I must mention that one of the best function machines around is the calculator. Use of the calculator in mathematics learning should begin in elementary school, and that is where the concept of function should be introduced. In that context, it should be patently clear that a function should be defined as a rule, or perhaps as a certain kind of machine, but certainly not as a set of ordered pairs!

Pedagogical Value

Some topics, which may or may not have intrinsic value, must be included in the curriculum because of their pedagogical value. By this I mean that some topics are important, not for their own utility, but rather because they form a necessary foundation or supporting structure for some other topic or topics that have intrinsic value.

As an example, consider the technique for completing the square. Although some might argue that this technique has intrinsic value (for example, in finding integrals of certain rational functions), the importance of the technique is difficult to argue for students who do not plan to study calculus. Even for those students who do plan to study calculus, the intrinsic value of completing the square is very limited. With the availability of symbol manipulation software, computers can calculate integrals of rational functions more rapidly and more accurately than any mathematician. The need to do such computations by hand is comparable to the need to do challenging long division problems by hand now that calculators are available. The need is not there!

Why, then, do we teach completing the square to algebra students? Because completing the square is a critical step in the derivation of the quadratic formula, and the quadratic formula is generally regarded as having intrinsic value.

Intrinsic Excitement

Some topics are just so interesting and exciting that their inclusion in the curriculum does not require any other justification. Examples abound in the sciences. Would anyone argue that a high school biology course should not contain some of the ideas of human genetics, that a chemistry course should not contain some information about laser chemistry, or that a physics course should not contain a discussion of elementary particles? These ideas, which are on the frontier of scientific research, are exciting to students and scientists alike. Teaching about these ideas communicates to students that the sciences are alive, fun to study, and worthy candidates for an interesting career.

In algebra, exponential growth and decay, especially if taught within the context of population dynamics or pollution control, can fit the criterion of intrinsic excitement. However, there is much more that is not now taught but could be, that would bring into the classroom not only the beauty and excitement of mathematics but also the flavor and vibrancy of current mathematics research.

For example, a student with access to a modest microcomputer and with a minimal understanding of quadratic functions and the complex plane can explore some of the ideas of fractals and chaos (see, e.g., Peitgen & Richter, 1986; Peterson, 1987). Chaotic dynamics is one of the most exciting areas of current mathematics research, and it is a source of some strikingly beautiful pictures. Students can create some of these pictures themselves, they can explore the consequences of small perturbations in parameters, and they can make and test conjectures.

There are other exciting areas of modern mathematics and its applications that can be at least discussed in secondary school mathematics classes. Tomography—the science of reconstructing images of the interior of an object, such as the human body, from shadow images, such as those obtained from CAT scans—is a mathematical science. Simulation—the technique by which the results of experiments that are too expensive, time consuming, dangerous, or otherwise impossible to carry out in practice are obtained theoretically using high speed computations on sophisticated computers— is a technique that not only uses mathematics but also stimulates the development of new mathematics.

Although these latter two examples are not algebra per se, they are nevertheless appropriate for discussion in an algebra class. As Steen (1986) has asked, "How many biology teachers would feel comfortable if they never mentioned DNA or viruses to their high school biology classes—simply because they were discovered after the teachers themselves were educated? Shouldn't mathematics teachers be just as embarrassed if they fail to introduce topics like tomography or simulation in their mathematics classes?" (p. 5).

Somehow, the fact that mathematics is a growing discipline, that current research in mathematics is yielding surprising, exciting, and valuable new results, that mathematics was not completed by the work of Euclid, or Newton, or Descartes—these facts must find their way into mathematics classrooms, from elementary school through high school and beyond. In particular, we must introduce more topics into our algebra classes on the basis of their intrinsic excitement.

EVALUATING STANDARD ALGEBRA TOPICS

The three criteria—intrinsic value, pedagogical value, and intrinsic excitement—can be used to measure the importance of including a given topic in a mathematics course. Let us examine from this perspective some of the topics that are standard fare in secondary school algebra courses.

The Real Number System

Surely there is no debating that some study of the real number system is necessary and appropriate, for both its intrinsic and its pedagogical value. High school algebra is, after all, primarily the study of polynomials (in one or several variables) with real number coefficients. The study of the solution sets of polynomial equations, even those with integer coefficients, requires a good understanding of real numbers, both rational and irrational. There is, however, room for debate on how much depth is optimal and which approach to representing the real numbers is most advantageous.

From the perspective of one who frequently teaches calculus to entering university students, I would prefer that students come to the university with a solid base of understanding about real number operations—including absolute value and exponentiation—and a good sense of numbers—positive and negative, rational and irrational. Moreover, students should be comfortable with inequalities, to the point that it is not a major task to determine which is larger:

$$-1.6 \text{ or } -2.4 \text{ ?}$$
$$3/8 \text{ or } 4/9 \text{ ?}$$
$$\sqrt{3} - 1 \text{ or } \sqrt{2} \text{ ?}$$

Instruction and exploration that lead to an appropriate level of understanding of real number operations is justified by intrinsic value. For example, familiarity with exponents is indispensable for comprehending very large numbers (like 2×10^9) and very small numbers (like 10^{-7}), as well as for understanding growth and decay—topics that are critically important for an informed citizenry and potentially exciting for students.

Regarding the representation of real numbers most appropriate for school use, I believe the time has come when we must adopt the decimal representation as primary. By this I mean that decimals should be taught in elementary school before fractions and, whenever fractions or radicals are used, the final form of the solution to any problem should always be expressed in decimal form, correct to a specified number of places. Consider the advantages.

First, the use of decimal representations would certainly enhance the development of number sense. For example, when numbers are expressed in decimal form, the order relations between them are transparent. Children would internalize number facts like "2/7 is less than 1/3, which is less than 3/8" ($.286 < .333 < .375$), much sooner than they do now.

I am not arguing that fractions and radicals should not be taught. To the contrary, I believe that both of these topics are essential (have intrinsic value). But to convey the impression to our students that numbers represented in radical form are meaningful and useful solutions to real-world problems is misleading.

A second advantage of using decimal representations is that addition and subtraction are more natural and straightforward. We have to address the problem of roundoff error, but that should be more comprehensible to students than finding least common denominators or rationalizing denominators.

For a start, we could truncate our decimals rather than rounding off. I believe that even very young children can understand that, with truncation, 3.14 represents a number that is 3.14 or larger but not as large as 3.15. It is then a small step to recognize that, when numbers are added, the truncated portions can accumulate so that perhaps 3.14 + 2.16 might be 5.31 rather than 5.30. Further, if we want to see which is correct, we can use three-place rather than two-place precision and then truncate. Or we can use a calculator, which will do the computation for us (recognizing, of course, that calculators do not truncate—they round off).

It is the wide availability of calculators that forms the basis for my third and, I believe, most compelling argument for treating the decimal representation as primary. Calculators are everywhere—in stores, in banks, in homes. The calculator is the tool that adults use to do arithmetic. Children know this. Many have their own calculators. Whether we approve or not, students will use calculators. And when they do, they will immediately encounter decimal representations. They will want to understand what these representations mean. Students should have the opportunity to learn about decimals, in school, in the early grades. Indeed, with calculators available, students will be highly motivated to learn, and to explore numbers. I firmly believe that calculators should be available for use in school mathematics whenever appropriate. And their use is appropriate whenever a student thinks their use would be helpful!

Relations and Functions

I have already commented on the importance of functions and my unhappiness with the approach to functions using ordered pairs. Those texts that use ordered pairs to define functions generally begin with a discussion of relations. Their treatment of relations is supposedly justified by pedagogical value: A function is a special case of a relation, so students must understand relations in order to understand functions. Wrong! The concept of relation is more abstract than the concept of function. Functions are actually very concrete objects. Let us not confuse our students by leading them to believe that the concept of function is difficult and that they cannot understand functions until they have understood relations. I can see no justification for teaching relations in secondary school algebra—not intrinsic value, not pedagogical value, and certainly not intrinsic excitement!

Let us, however, teach functions to a much greater extent than we do now. Let functions permeate the course. I believe that functions should form the backbone of a first course in algebra. Intrinsic value? Certainly. Pedagogical value? Yes. Intrinsic excitement? Yes, indeed.

I commented earlier on the intrinsic value of the function concept. Let me now comment briefly on the pedagogical value and intrinsic excitement.

Secondary school algebra is primarily the study of polynomials—what they are; how to add, subtract, multiply, and divide them; how to find their "zeros"[2]; how to use them to solve problems. What is a polynomial? The favorite definition seems to be that a *polynomial* is a sum of monomials, where a *monomial* is defined to be:

- a numeral or constant, a variable to a power, or a product of a numeral and a variable to a power [AW],

- a term that is the product of numbers and variables with nonnegative integral exponents [HBJ], or

- a term which is either a numeral or a variable, or a product of a numeral and one or more variables, raised to positive powers [HM].

Aside from the fact that two of these definitions may be wrong (in allowing fractional powers), these are extraordinarily cumbersome definitions for the primary object of study in a beginning algebra course.

To be sure, it is essentially this sort of definition, phrased with a bit more clarity, that is used in more advanced abstract algebra courses to define polynomials over arbitrary fields. But for the study of polynomials with real number coefficients, it is sufficient to define a polynomial to be a special kind of function, one that can be expressed as a sum of functions of the form $f(x) = ax^n$ (where a is a real number and n is a nonnegative integer). For a student who already has some experience with functions, this definition is quite natural. It helps unravel the mystery surrounding problems like "simplify the following expressions." (What is the meaning of these "expressions" anyway? Why should we want to "simplify" them?) The task of finding a simpler way to express a given function is one whose benefits are clear.

So I advocate a function approach to algebra. I believe that, from the function perspective, most of the manipulative "games" that students are required to play in an algebra course will become more meaningful.

At the same time, perhaps we can eliminate the unfortunate use of the word *expression*. Asking students in an algebra course to manipulate expressions is analogous to asking students in a writing course to manipulate phrases rather than sentences. Expressions are not important in themselves. They are important only when they are implicitly or explicitly part of an equation. The expression $2x + 1$, by itself, is incomplete. To have meaning, it must be imbedded in an equation, such as $f(x) = 2x + 1$, or $2x + 1 = 0$. The equation provides meaning for the expression, as well as a context for x. (Is x a variable or does x represent a member of a solution set?) Just as we teach students of writing to speak in sentences, let us teach students of algebra to speak in sentences![3]

Returning to the value of teaching functions in algebra, I have spoken of the intrinsic value and the pedagogical value. Is there also intrinsic excitement? I believe so. At least, there can be. Let me give just one example. As I have said earlier, the calculator is a wonderful function machine. It is a fascinating exercise to iterate functions with a calculator and see what happens. For example, iterating the square root function, the sine, the cosine, and the exponential function reveals four very different and interesting behaviors.

Functions are basic—in algebra, in mathematics. Let us not just teach about functions in algebra; let us make functions the centerpiece of algebra instruction.

Graphs

Another topic of clear intrinsic importance is the topic of graphs—bar graphs and pie charts, as well as graphs of functions of one and two variables. An understanding of graphs is required to appreciate fully the weather report on television or in the newspaper and to understand trends in such diverse areas as air quality and stock market performance. Functions and graphs should go hand in hand, each reinforcing the understanding of the other.

There is no need to restrict attention in an algebra course to polynomial and rational functions, however. With graphic calculators and computers available, students can (and should!) explore graphs of exponential, logarithmic, and trigonometric functions as well. A nice discussion of the use of computers in studying graphs can be found in Small (1986).

Algebraic Manipulations

Whether it is the manipulation of algebraic expressions or the manipulation of algebraic equations, there is room for discussion and research on how proficient students in today's technological world need to be in manipulative skills. Just as the widespread availability of calculators has removed much of the rationale for insisting on a high level of competence in the algorithms of arithmetic, the growing availability of microcomputers and symbol manipulation software is about to remove much of the rationale for insisting that algebra students attain a high level of competence in symbol manipulation.

Existing software packages will simplify algebraic expressions; add, subtract, multiply, and divide polynomials and polynomial fractions; factor polynomials; differentiate and integrate; find solutions of equations; and graph functions—all in response to a few simple keystrokes. True, these packages are expensive and often not very user friendly, but the prices are coming down and the friendliness of the software is improving. Clearly, in the not-too-distant future, all serious users of algebra will do their calculations on computers, just as all serious users of arithmetic rely on calculators now. That time is not far away. Already there is a sophisticated scientific calculator (the HP-28C) that will do these tasks, and it is available for less than $180.

Henry Pollak, writing for the Conference Board of the Mathematical Sciences, aptly described the challenge before us:

> The basic thrust in Algebra I and II has been to give students moderate technical facility. When given a problem situation, they should recognize what basic algebraic forms they have and know how to transform them into other forms which might yield more information. In the future, students (and adults) may not have to do much algebraic manipulation—software like mu-Math will do it for them—but they will still need to recognize which forms they have and which they want. They will also need to understand something about why algebraic manipulation works, the logic behind it. In the past, such recognition skills and conceptual understanding have been learned as a by-product of manipulative drill, if learned at all. The challenge now is to teach skills and understanding even better while using the power of machines to avoid large time allotments to tedious drill. (Conference Board of the Mathematical Sciences, 1983, p. 9)

Just as the goal of arithmetic instruction needs to shift away from skill with complex arithmetic operations and toward skill with estimation, mental arithmetic, and approximation, so also must the goal of algebra instruction shift away from skill with algebraic manipulation and toward conceptual understanding and problem-solving skills.[4]

Word Problems

Word problems have the potential of being intrinsically important, pedagogically important, and intrinsically exciting—all three. But, too often, word problems that appear in algebra textbooks are contrived. For example:

- Tim is twice as old as Art. Tim is also exactly 15 years older than Art was last year. What are their ages? [HM, p. 130]
- Find two consecutive integers whose product is 110. [HBJ, Vol. 1, p. 340]
- The hypotenuse of a right triangle is 25 feet long. One leg is 17 feet longer than the other. Find the lengths of the legs. [AW, p. 495]

None of these problems illustrates a real-world application of algebra. They are "puzzle problems." Possibly they are fun to do. Certainly they do lead to algebraic equations to solve. But the frequency with which such problems appear in algebra texts misleads students about the value of algebra. As Usiskin (1980) has pointed out, current word problems, though intended to teach translation of real-world conditions into mathematical form, are so contrived that they succeed only in teaching students that algebra "has no applications, that you should not use common-sense arithmetic to solve [problems], and that algebra is difficult" (p. 416). Worse yet, students are too often drilled in special techniques for solving each class of problems.

Now, I do not mean to imply that we should eliminate word problems from algebra. Let us have lots of word problems in our algebra texts! Indeed, let us constantly motivate the study of algebra by such problems. But let them be problems with meaning, problems that youngsters can see as being important, problems that students themselves may need to solve sometime. Most books do have some problems of this type, but there are not enough of them, and they do not occupy a position of prominence. Let us feature these problems and relegate traditional algebra word problems to a "puzzle section" of the problem set, so it is clear that they are, after all, only puzzles and of very little intrinsic importance.

Quadratic Formula Versus Numerical Methods

In view of the available technology, can we still justify teaching the quadratic formula? From the perspective of intrinsic value, I think not. The quadratic formula does provide a straightforward method for finding the solutions of quadratic equations in closed form. But which are the more meaningful solutions of the equation

$$x^2 + 2x - 5 = 0,$$

$-1 + \sqrt{6}$ and $-1 - \sqrt{6}$, or 1.449 and -3.449 (three-place accuracy understood)? There's no contest! And what is the easiest method for obtaining these solutions? Not by using the quadratic formula and then using a calculator to convert the radicals to decimals. Use the solve key! Or, if the calculator does not have a solve key, use the bisection method or some other rapidly converging algorithmic method.

Should students learn how the calculator got the solutions when the solve key was pressed? Perhaps. They should certainly learn something about numerical methods. Numerical techniques for solving equations are important, but not because these are the techniques that machines use and not because of their value in solving quadratic equations, but because:

> First, . . . long after students have forgotten the quadratic formula or developed bugs in their factoring skills, they will likely know enough to try successive approximation. Second, the method of successive approximations applies without a major change to the solution of nearly every type of equation, from higher-degree polynomials to rational, exponential, or trigonometric conditions. Third, the numerical approaches stress the functional point of view that is at the heart of most later applications of algebra. (Fey, 1984, p. 15)

Can we justify teaching the quadratic formula for its pedagogical value? I think so, but not because it leads to useful and efficient methods for solving equations. The quadratic formula is too special. It applies only to quadratics; it does not generalize. Although we can, with some effort, solve cubics and quartics by radicals, it is not possible to solve quintic and higher-degree polynomial equations that way. Algorithmic methods, on the other hand, are completely general.

The pedagogical value of the quadratic formula is that it leads naturally to complex numbers, and complex numbers are intrinsically important. Using the quadratic formula, we can see that every quadratic equation can be solved provided that we extend our number system appropriately. And the resulting number system is the most important number system in mathematics.

The Complex Number System

Complex numbers are of fundamental importance in a variety of fields, including physics and electrical engineering. The first time that students begin to encounter these applications seriously is probably in a course in differential equations. Nevertheless, there is an opportunity to discuss complex numbers in secondary school algebra, and that opportunity should not be missed. Complex numbers can be discussed from a historical perspective. Some preliminary idea of how complex numbers are used in physics and engineering can be shared with students. And, as I have mentioned earlier, with the help of a microcomputer, students can explore chaotic subsets of the complex plane and, in the process, get involved in one of the most exciting and beautiful areas of current mathematics research.

Tables and Interpolation

All of the algebra textbooks I have examined include tables in the back— tables of square roots and tables of trigonometric functions. It seems to me that the time has come to delete these tables from textbooks. Calculators can provide all the relevant numbers more quickly and more accurately.

I am not arguing that tables are unimportant—just that these particular tables are no longer useful. It is important that our students understand tables in general. Tables are all around us—in newspapers (sports, financial pages), in books (health, nutrition), even on income tax forms. Students should be comfortable with tables, and they should work with tables in an algebra course if they have not done so earlier.

A good way to understand tables is to construct some of them. With a calculator, for example, students can construct tables of trigonometric functions without a great deal of effort, and they will learn some of the characteristics of these functions in the process.

Although I did find tables in the algebra texts I examined, I failed to find any discussion of interpolation. Is interpolation a lost art? Its importance can be appreciated by anyone who rides city buses and needs to estimate arrival times at a stop that is intermediate to those that are listed on the bus schedule. Linear interpolation is a nice application of linear functions and is therefore appropriate for inclusion in an algebra course. Moreover, linear approximation is a topic that students who continue their study of mathematics will see over and over again in more advanced courses.

But I would not stop with linear interpolation. Polynomial interpolation and polynomial approximations are very important in applied mathematics. There is absolutely no reason to deprive high school students of the opportunity to study these topics. Polynomial interpolation and approximation provide much more realistic applications of algebra than most of the "applications" that one finds in typical algebra textbooks. And, with the help of a microcomputer, students can readily explore the relevant ideas.

Systems of Linear Equations and Linear Algebra

Systems of linear equations are important in the social sciences, especially in economics. All of the books I examined include a discussion of systems of linear equations, but most of them restrict attention to two equations and two variables. I believe the treatment of linear systems should be expanded to cover m equations in n variables. The techniques for solution, if cast in matrix form, are not difficult. And only by discussing larger systems of equations can one meaningfully discuss the important concepts of underdetermined and overdetermined systems.

Perhaps a full course in elementary linear algebra, similar to courses that are taught to freshmen and sophomores at many colleges and universities, should be an option for Grade 12. Linear algebra is accessible to any student who has mastered Algebra I, and it is useful in the physical sciences, as well as in the social sciences. Moreover, it can be taught with a geometric emphasis, which would enhance students' understanding of translations, rotations, and reflections, in both two- and three-space (see, e.g., Thorpe & Kumpel,

1984). Thus, a course in linear algebra could help to correct one of the failings of the current secondary school curriculum—the nearly total lack of attention to three-dimensional geometry and spatial perception.

SUMMARY

I stated at the beginning of this paper that the teaching of algebra in the schools has not changed significantly in fifty years. This is really not surprising. The content of algebra has not changed as dramatically as the content of biology, chemistry, and physics. Our students have needed to learn pretty much the same algebra content as did their parents and grandparents.

But now something *has* changed. We have new tools. Calculators and computers are widely available, and they are becoming more and more sophisticated even as their cost comes steadily down. It would be foolhardy to ignore the potential for using these tools to revolutionize the teaching of algebra. In the use of algebra by professionals, the revolution has already begun.

FOOTNOTES

1. Actually, Flanders was discussing goals for calculus instruction, but this goal applies just as well to algebra.
2. The use of the word *zero* in this context seems to have caught on in school algebra. I, for one, cannot get used to it, and I do not think that its use here is pedagogically sound. The concept of zero as a real number is already sufficiently mysterious without compounding the problem by using the same word to describe an entirely different concept. I marvel when anyone seems to be comfortable making a statement like "Two is a zero of $x^2 - 4$."
3. I notice that some of the Ohio State Middle School Mathematics Project materials (field test version, 1985-6) use the words *mathematical phrase* in place of *expression*. This is a step forward—it calls a spade a spade—but let us move one more step. Let us always imbed our mathematical phrases in complete sentences!
4. The list in Fey (1984, pp. 21-22) of manipulation topics that should or should not be taught, based on intrinsic value, is a good start.

REFERENCES

Conference Board of the Mathematical Sciences. (1983). The mathematical sciences curriculum K-12: What is still fundamental and what is not? In *Educating Americans for the 21st century* (Source Materials, pp. 1-23). Washington, DC: National Science Board Commission on Precollege Education in Mathematics, Science and Technology.

Coxford, A. F., & Payne, J. N. (1987). *HBJ Algebra 1 & 2*. Orlando, FL: Harcourt Brace Jovanovich.

Dolciani, M. P., Wooton, W., & Beckenbach, E. F. (1983). *Algebra 1*. Boston, MA: Houghton Mifflin.

Fey, J. T. (Ed.). (1984). *Computing and mathematics: The impact on secondary school mathematics*. Reston, VA: National Council of Teachers of Mathematics.

Keedy, M. L., Bittinger, M. L., Smith, S. A., & Orfan, L. J. (1984). *Algebra*. Menlo Park, CA: Addison-Wesley.

Peitgen, H. O., & Richter, P. H. (1986). *The beauty of fractals*. New York: Springer-Verlag.

Peterson, I. (1987). Zeroing in on chaos. *Science News, 131*, 137-139.

Small, D. (1986). Computer algebra systems in undergraduate instruction. *College Mathematics Journal, 17*, 423-433.

Steen, L. A. (1986, November). *Forces for change in the mathematics curriculum.* Paper presented at "The School Mathematics Curriculum: Raising National Expectations," conference sponsored by the Mathematical Sciences Education Board and the UCLA Center for Academic Interinstitutional Programs, Los Angeles, CA.

Thorpe, J. A., & Kumpel, P. G. (1984). *Elementary linear algebra.* Philadelphia, PA: Saunders College Publishing.

Usiskin, Z. (1980). What should *not* be in the algebra and geometry curricula of average college-bound students? *Mathematics Teacher, 73*, 413-424.

Whitney, H. (1985). Taking responsibility in school mathematics. *Journal of Mathematical Behavior, 4*, 219-235.

Critical Considerations for the Future of Algebra Instruction
or
A Reaction to: "Algebra: What Should We Teach and How Should We Teach It?"

Joan R. Leitzel
Department of Mathematics
The Ohio State University

John Thorpe (this volume) has brought into focus several important issues with respect to the content and methodology of algebra in the schools and has set a high tone for our discussions. To a large extent, he and I share the same attitudes and agree on issues related to curriculum. In that sense, my response may fail to identify some issues that others feel are controversial. Nevertheless, I would like to highlight some of John's points that I think are particularly pivotal in our deliberations, pose one more topic for discussion, and finally, fuss about one point where John and I perhaps disagree.

FOUR KEY POINTS

Reconsidering Content and Approach

John made the point that the teaching of algebra in the schools has not changed significantly in at least fifty years, even though mathematics has changed dramatically. To test that assertion, I compared two texts on my shelf—one with copyright 1934, the other printed in 1932—with the elementary algebra text that my son used in eighth grade, only five years ago (Dolciani, Wooton, Beckenbach, Jurgensen, & Donnelly, 1967; Hart, 1934; Schultze, 1905/1932). Indeed, the mathematics content in these books does not differ significantly. The older books do more theory of equations and mathematics of finance than the newer one. Also, we have gotten much more wordy—the newer text is approximately twice the length of the older ones. And, the approach to the mathematics is noticeably different in the newer text.

Let me give an example of the change in approach: In the 1932 book, in Chapter 6, entitled "Multiplication," the author persuades students that the product of two negative numbers is positive by telling a story about two tramps who came to town and cost the town $3.00 a day each. When the tramps arrived, the cost to the town was represented as 2×-3 or -6. When they left town, the saving was represented by -2×-3 or 6. In contrast, the modern text begins with axioms for the real number system. Then, on page 83 we have this argument that $a(-1) = -a$:

/3 7, / 5⁻8

1. a is a real number   Hypothesis
2. $a(-1)$ is a real number   Axiom of closure for multiplication
3. $a = a \times 1$ Multiplicative axiom of one
4. $a + a(-1) = a \times 1 + a(-1)$ Additive property of equality
5. $= a[1 + (-1)]$ Distributive axiom
6. $= a \times 0$ Axiom of additive inverses
7. $= 0$ Multiplicative property of zero
8. But $0 = a + (-a)$ Axiom of additive inverses
9. $a + a(-1) = a + (-a)$ Transitive property of equality
10. $a(-1) = -a$ Cancellation law for addition

This example suggests a further reason why it is imperative that we reconsider both the content and the approach to algebra in the school curriculum: Large numbers of students are not learning the algebra they study. In autumn quarter 1985, of the 6430 freshmen who entered Ohio State University, 10% had less than three years of college preparatory mathematics, 66% had more than three years, and 24% had exactly three years, including two years of algebra. In this last group, almost half of the students were placed in remedial mathematics on the basis of their placement test scores; more than a quarter of these students were not able to use a variable in any context. These students demonstrated no ability to use elementary algebra even though they had credit for two years of high school algebra.

We need to change the approach to teaching algebra for at least two reasons: first, to make essential concepts more understandable to students; second, to give students access to topics that otherwise would be delayed until late in the curriculum.

New Computational Tools

The availability of calculators and computers permits a numerical approach to algebra quite different from the formal axiomatic approach of recent years. Although computers may have their day in the future, especially in the area of graphics, for the present it is the hand-held calculator that has the potential for greatest impact on the teaching of algebra and other mathematics. The low cost of sophisticated scientific calculators means that every student, from middle grades on up, can be assumed to have a calculator. Computation can therefore be used intentionally to introduce key concepts of algebra.

Algebra is an extension and a generalization of arithmetic. It provides tools for solving problems that arithmetic does not provide. Algebra did not develop historically as an axiomatic structure. Algebraic structure was investigated late and, as useful as it is in mathematics, it is too sophisticated an approach for beginning students.

The advantage of having calculators in algebra courses is that students can start with problems and calculators, rather than starting with axioms. Problems, rather than formalism, can be used to introduce students to key concepts. Formal processes can be delayed. Students can solve problems without resorting to formulas and mechanical processes. Graphing can be introduced early and can provide a concrete representation of many algebraic ideas. Students can even graph using certain calculator function keys.

Of course, a calculator is not essential for taking a problem-solving approach. I was fascinated to see that in my 1932 book, each chapter begins with a problem for student investigation. In Chapter 1, "How Letters Are Used as Numbers," the introductory problem is to find the height of the classroom by holding a right isosceles triangle at eye-level and sighting along the hypotenuse. Chapter 2 of this text is called "Graphs." Although the vocabulary and the examples are from another day, the approach in this chapter is sound. Problems are used to introduce students to key concepts, and concrete representations are used when they can be found. Many of us, including John, might feel more comfortable with this book than one of the modern texts. (At least there is no talk about sets of ordered pairs or any statement like "Two is a zero.") Still, this 1930s book could not assume that students can develop mathematical ideas through extensive computation, and this is a limiting feature.

Importance of Numerical Methods

In elementary algebra we pretend that all equations are either linear or quadratic. We convert students into equation-solving machines. Asked to find a value for x that satisfies the equation $5x = 10$, students go through the steps mechanically and write down things like this:

$$5x = 10$$
$$\frac{5x}{5} = \frac{10}{5} \quad \text{Divide both sides by 5}$$
$$x = 2$$

Further, we imply that every solution can be expressed in closed form, and we choose only problems that can be represented by equations that students have mechanical techniques to solve. No wonder the courses are dull!

On the other hand, a student with a calculator can readily solve an equation of this form:

$$500 \left(1 + \frac{.075}{12}\right)^x = 1000$$

A student who is accustomed to reasoning numerically observes that the solution to this equation is the value of x that makes $(1 + .075/12)^x$ as close to 2 as possible. On a calculator—especially, a calculator with a constant key—that value is quickly found.

As John has observed, young students can investigate problems of compound interest, or population growth, or radioactive decay. They can understand what it means for an equation to have a solution before they learn formal procedures for finding a solution. What is more, students get very good at finding solutions to equations by approximating and refining their estimates—so good, in fact, that they sometimes wonder why teachers spend so much time on formal methods.

The Concept of Function

John has pointed out that functions are the "backbone of algebra." We are well reminded of the importance of this concept. The function idea is so basic that students should experience it very early in their mathematics training. Consider John's example of sales tax. Long before students can represent relations algebraically, they should understand that sales tax is a function of price (see Figure 1).

Rather than ask a young student, "What is the sales tax on a refrigerator that costs $800.00?" we can ask the student to compute the sales tax on a wide range of appliances so that the student develops intuition about the relationship between the cost of the item and the amount of sales tax.

A table is a concrete representation of a function. A table is a set of ordered pairs, but I agree with John, we do not need to use that vocabulary in school algebra. Tables, in turn, can be represented by graphs before students are introduced to algebraic notation (see Figure 2).

A graph is also a set of ordered pairs, but more importantly, a graph is another concrete representation of a function—a visual tool that every calculus teacher wishes students used naturally. Throughout the secondary school curriculum, I would argue that algebra should have a strong geometric flavor.

CONCEPT OF VARIABLE

After considerable numerical experience, a student is ready for the alge-

Marked Price ($)	Sales Tax at 5.5% ($)	Total Cost of Item ($)
1.00	1.00 x .055 = .06	1.00 + .06 = 1.06
5.50	5.50 x .055 = .30	5.50 + .30 = 5.80
189.64	189.64 x .055 = 10.43	189.64 + 10.43 = 200.07
880.00	880.00 x .055 = 48.40	880.00 + 48.40 = 928.40
1000.00	1000.00 x .055 = 55.00	1000.00 + 55.00 = 1055.00

Figure 1. Sales tax as a function of marked price.

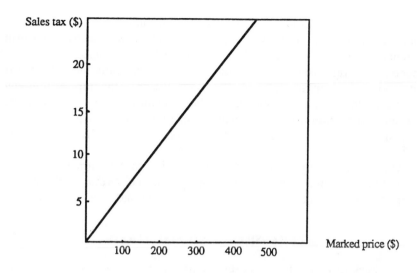

Figure 2. Graphical representation of sales tax as a function of marked price.

Marked Price ($)	Total Cost of Item ($)
1.00	1.00 + 1.00 x .055
5.50	5.50 + 5.50 x .055
189.64	189.64 + 189.64 x .055
880.00	880.00 + 880.00 x .055
1000.00	1000.00 + 1000.00 x .055
P	P + P x .055

Figure 3. Total cost as a function of marked price.

braic representation of a function. In the sales tax example, students who have computed the sales tax and total cost for several items should be able to generalize the procedure and represent the total cost of any item in terms of a variable that stands for the marked price (see Figure 3).

Variables are very basic in algebra. They provide the algebraic tool for expressing generalizations. But, the concept of variable is more sophisticated than we often recognize and frequently turns out to be the concept that blocks students' success in algebra. Students need to become comfortable with variables in numerical contexts before they begin a formal study of algebra. I would like to add *variables* to our discussion of those features of content and approach that need greater attention.

To many students, using a letter to represent numbers or other mathematical objects is completely mysterious. Thus, we must pay careful attention to how this concept is introduced and what students' early experiences with variables are. Often the first experience in using a variable is to represent an unknown number in an equation. This introduction can limit a student's understanding of variables. A student who views a variable merely as an unknown number may in time have difficulty understanding sentences that begin, "For all real numbers, x, . . ." or "Let v be in the kernel . . . ."

When students introduce variables into tables, to summarize in one line several lines of numerical data, they come to appreciate that a variable can represent objects in a large set—a very important understanding indeed.

EXPRESSIONS AND EQUATIONS

Now back to John's critical observations about functions. What do we do about this unfortunate word *expression*? John has mentioned my efforts to talk about a mathematical phrase, rather than a mathematical expression. Our intention in writing materials for middle grade students has been, where possible, to use language that is meaningful outside of mathematics. However, I am sorry to report to you that when those materials were edited by a publisher, the word *phrase* in each instance was changed to *expression*. The explanation for the change was this: There is no such thing as a mathematical phrase. So there!

The point I would like to fuss about is John's statement that mathematical expressions, or phrases, or whatever, are important only when they are part of an equation. I would argue that, before students see algebraic equations, they need to have considerable experience with mathematical expressions arising from concrete problem situations. In the example seen earlier, the expression $P + P \times .055$ arose naturally, not in an equation, but as an important generalization of a certain set of numbers. It has meaning in that context. Furthermore, when an expression arises from a concrete setting, one in which a student needs to do considerable computation, the student sees utility in the distributive property that permits $P + P \times .055$ to be replaced by $1.055P$. Students are motivated to simplify expressions when simplification saves time in numerical computation.

When algebraic expressions arise in concrete problem settings as a means of summarizing numerical information, meaningful equations are suggested, and students can gain experience in identifying solutions to equations before they learn mechanical ways of solving equations. For example, beginning students would see that the marked price of an item whose total cost is $928.40 is $880. That is:

$$\text{If } P + P \times .055 = 928.40, \text{ then } P = 880.$$

Graphs also are meaningful to students before equations are introduced. In our simple example, the marked price, P, can be associated with the total cost, $P + P \times .055$, to form a graph. This graph is meaningful before anyone says, "Consider the equation $y = P + P \times .055$, or the function $f(P) = P + P \times .055$."

Students tend to confuse the concept of function and the concept of equation. Last year I sat in on a Ph.D. oral examination in mathematics education and asked the student this question: Do *function* and *equation* describe the same mathematical concept or are they conceptually different? Of course, every Ph.D. candidate says at least one crazy thing in the oral exam, but this student set a new record by explaining that a function is an equation and an equation is a function. We need to be sure that those ideas do not get muddled in young students' heads. Therefore, I say delay the introduction of equations and, especially, delay the introduction of functional notation until students have had concrete experience with functions and have learned several ways of representing functions.

ADDITIONAL IMPORTANT POINTS

In closing, let me mention quickly four other points John made that I expect to be particularly fruitful in our deliberations:

1. We must set aside the notion that the gateway to mathematics is through the rational numbers. We are sacrificing thousands and thousands of students every year on the altar of fractions when they could be involved with meaningful mathematics.

2. As John has suggested, linear algebra—perhaps linear algebra with statistics—can have an important role as an alternative to calculus in Grade 12. Too many students are taking 12th-grade calculus and repeating it again in college. Their time would be better spent on other mathematics. Linear algebra can be very accessible to young students; it has a clear geometric basis; interesting and important problems abound. Plus, linear algebra can serve to consolidate a student's understanding and skills in algebra.

3. Complex numbers have intrinsic value in mathematics. Opportunities for students to have experience with complex numbers should not be overlooked in the secondary curriculum. Geometric representation of the complex numbers and the operations on complex numbers is particularly important. In addition, young students can understand the arithmetic similarities among Gaussian integers, rational integers, and polynomials— therein gaining an introduction to algebraic structure, plus a clearer understanding of factorization into primes.

4. Finally, John has stated the appropriate goals for algebra: conceptual understanding and problem-solving skills. He has given three useful criteria for selecting topics to meet these goals. Both his criteria and his examples

are excellent. I personally would like to see more topics qualify on the criterion of intrinsic excitement. Because students often find mathematics uninteresting, making progress on this criterion will require attention to teacher education. Teachers of mathematics must be constantly on the lookout for examples, for new developments in mathematics, for applications that make sense to students. Although textbooks can provide some help with this, they cannot substitute for a teacher who is alert to new developments. Thus, mathematics departments in universities and colleges must assume greater responsibility in providing for the continuing education of teachers of mathematics.

REFERENCES

Dolciani, M. P., Wooton, W., Beckenbach, E. F., Jurgensen, R. C., & Donnelly, A. J. (1967). *Modern school mathematics: Algebra 1*. Boston, MA: Houghton Mifflin.

Hart, W. W. (1934). *Progressive first algebra*. Boston, MA: D. C. Heath.

Schultze, A. (1932). *Elementary and intermediate algebra*. New York: Macmillan. (Original work published 1905)

The Early Learning of Algebra: A Structural Perspective

Carolyn Kieran
Département de Mathématiques et d'Informatique
Université du Québec à Montréal

Algebra is an area of human problem solving that has been of interest to researchers for several decades. Recently, however, this interest has escalated dramatically. Because of this new surge, the time now seems right for us to synthesize what we know of the learning of algebra in an effort to suggest future research directions.

This paper attempts to provide part of that synthesis by reviewing the research on the initial learning of algebra. The teaching of high school algebra usually begins with the following topics: variables, simplification of algebraic expressions, equations in one unknown, and equation solving. Students' difficulties with these topics have been found to center on (a) the meaning of letters, (b) the shift to a set of conventions different from those used in arithmetic, and (c) the recognition and use of structure. The review presented in this paper includes studies related to these three areas of difficulty but emphasizes the last one—recognition and use of structure—for this is the core of high school algebra. Since school algebra can be considered as the formulation and manipulation of general statements about numbers, children's prior experience with the structure of numerical expressions in elementary school should have an important effect on their ability to make sense of algebra. Thus, this paper begins with a section on pertinent research from the domain of elementary school mathematics.

ROOTS OF ALGEBRA IN ARITHMETIC

Much of elementary school arithmetic is oriented towards "finding the answer." However, there are certain activities that, in theory, emphasize some of the structural aspects of arithmetic, such as the representation of word problems by equations, the solving of open sentences (e.g., "missing addend" sentences), and the simplification and comparison of complex arithmetic expressions. Before examining children's recognition and use of structure in these situations, let me briefly characterize the ways in which the term *structure* is used throughout the paper.

Structure in School Algebra

The use of the phrase *arithmetic/algebraic structure* refers, in general, to the structure of a system that is comprised of a set of numbers/numerical variables, some operation(s), and the properties of the operation(s). But

33

this paper also includes references to structure in a more particular sense—to the structure of certain mathematical objects, such as the structure of expressions, the structure of equations, and the structure of word problems. *Structure* is defined in Webster's *New Collegiate Dictionary* to mean "the arrangement of the parts in a whole, the aggregate of elements of an entity in their relationships to each other." The former deals with arrangement or disposition; the latter with relationships. When referring to the *structure of an algebraic or arithmetic expression*, I mean to include both of these components, which I have named "surface" structure and "systemic" structure. More precisely, the *surface structure* of an algebraic or arithmetic expression refers to the given form or arrangement of the terms and operations, subject—when arranged sequentially—to the constraints of the order of operations. The *systemic structure* of an algebraic or arithmetic expression (systemic in the sense of relating to the mathematical *system* from which the expression inherits its properties) refers to the properties of the operations, such as commutativity and associativity, and to the relationships between the operations, such as distributivity. The systemic structure of algebraic expressions permits us to express, for example, $3(x + 2) + 5$ equivalently as $5 + 3(x + 2)$ or as $3x + 11$. Thus the structure of the expression $3(x + 2) + 5$ includes the surface structure, that is, the given ensemble of terms and operations—in this case, the multiplication of 3 by $x + 2$, followed by the addition of 5—along with the systemic structure, that is, the equivalent forms of the expression according to the properties of the given operations.

The *structure of an equation* incorporates the characteristics of the structure of expressions, for an equation relates two expressions. Thus the surface structure of an equation comprises the given terms and operations of the left- and right-hand expressions, as well as the equal sign denoting the equality of the two expressions. Similarly, the systemic structure of an equation includes the equivalent forms of the two given expressions. For example, the equation $3(x + 2) + 5 = 4x/2 - 7$ can be re-expressed as $3x + 11 = 2x - 7$, wherein each expression is independently transformed (i.e., simplified). Because of the equality relation inherent in an equation, the left-hand expression continues to be equivalent to the right-hand expression after such systemic transformations of one or both expressions. The resulting equation is also equivalent to the given equation.

However, the systemic structure of an equation includes much more than the systemic structure of expressions. Because of the equality relation and system properties such as the addition property of equality ("if equals are added to equals, the sums are equal"), the equation as a whole can be transformed into any one of a number of equivalent equations without necessarily replacing one or both expressions by equivalent ones. For example, the equation $3x + 11 = 2x - 7$ is equivalent to the equation $3x + 11 + 7 = 2x - 7 + 7$, even though the left-hand expression $3x + 11$ is not equivalent to $3x + 11 + 7$, nor is the right-hand expression $2x - 7$ equivalent to $2x - 7 + 7$. Similarly, the equation $5x + 6 = 10$ is equivalent to $5x = 10 - 6$, according to the structural properties of our arithmetic/ algebraic system, wherein an addition can be expressed equivalently as a subtraction (e.g., as in $4 + 3 = 7$ being expressed as $4 = 7 - 3$).

The system properties of equality can be used to generate an infinite set of equations, in fact, a class of equivalent equations. It is this particular aspect of the systemic structure of equations—that is, the potential of generating equivalent equations by means of properties related to (a) performing the same operation on both sides of an equation and (b) the alternate way of expressing additions and multiplications in terms of subtractions and divisions—that is so crucial to the process of solving equations.

Another usage of the term *structure* in this paper concerns the *structure of word problems*. To illustrate the sense in which this expression is used, I refer to a work of Vergnaud (1982) in which he set up a classification system for addition and subtraction word problems. The system that Vergnaud imposed upon this field of word problems comprised six main categories of relationships. Each category was further divided into subclasses. Thus, according to Vergnaud's classification, problems belonging to the same subclass are considered to be structurally equivalent. Structural equivalence is based on problem types rather than problem solutions. This can be illustrated by looking at two problems, each belonging to a different subclass of the same category:

- Fred had 12 sweets. He ate 5 of them. How many sweets does he have now?

- Tony had 5 marbles. He played marbles with Robert. Now Tony has 12 marbles. What happened during the game?

Though the same operation (i.e., $12 - 5$) may be used to solve both problems, the problem situations are different. The structure of the first problem is represented by $12 - 5 = \square$; the structure of the second by $5 + \square = 12$. In addition to illustrating what is meant by two different problem structures, these examples also show that representing the structure of a problem can involve an arithmetic operation that is not the same as the operation used to solve the problem. It should be noted that Vergnaud's structuring of classes of word problems is not unique; others such as Carpenter and Moser (1982) have structured the field differently. In keeping with Vergnaud's approach, however, whenever I speak of the structure of a word problem throughout this paper, I am using the term to refer to the relationships involved in the problem situation and not to the numerical operations performed to find the solution. Let us now examine some studies dealing with elementary schoolchildren's recognition and use of structure.

Equations and Arithmetic Word Problems

Researchers often distinguish algebra word problems from arithmetic word problems (Mayer, 1980). Though the distinction between the two types of word problems is fuzzy and may even be nonexistent, it is generally considered that, put simply, algebra word problems are those that are found in algebra textbooks and arithmetic word problems are those that are found in arithmetic books. The following is a typical arithmetic word problem:

> Joe had 4 marbles. Tom gave him some more. Then Joe had 7 marbles. How many marbles did Tom give him?

A typical algebra word problem is:

> Ben is 4 years older than Juan. In 2 years, their ages will total 50. How old is each now?

It has been suggested that for arithmetic word problems the child's mental representation of the problem specifies the operation to be carried out (Kintsch & Greeno, 1985), but that for algebra word problems it is not the solving operation but the problem situation that must be represented mentally. Because of the complexity of many algebra word problems, representing the problem situation generally requires the use of some written form of representation, such as an algebraic equation. However, even this distinction between algebraic and arithmetic word problems is limited. Many problems found in the arithmetic texts of Grades 4, 5, and 6 are too complicated to be solved mentally by children, and many so-called algebra problems can be solved without any algebra at all.

Though both of the problems referred to above can be represented by equations, children hardly ever use equations to represent arithmetic problems (Carpenter & Moser, 1982; Nesher, Greeno, & Riley, 1982; Vergnaud, 1982). In fact, if an equation in the form of an open sentence (e.g., $4 + ? = 7$) is requested, children will solve the problem first and then attempt to provide the equation afterwards (Briars & Larkin, 1984). Furthermore, it has been found that children who can solve word problems often cannot write equations to represent the quantitative relationships in the problem situations (Lindvall & Ibarra, 1978; Riley & Greeno, 1978).

In fact, there is a tension between representing the structure of the problem situation (for the above arithmetic problem: $4 + \square = 7$) and representing the procedure that is generally used to solve the problem ($7 - 4 = \square$). If children do write an equation, it usually represents the operations they carry out in arriving at the final answer to the problem ($7 - 4 = 3$). This often results in a statement that has an equal sign but lacks equivalence between the left and right sides of the equation. The following example illustrates the difficulty children experience in representing the structure of a problem situation. Vergnaud and his colleagues (Vergnaud,

Benhadj, & Dussouet, 1979) presented sixth-grade children with the problem:

> In an existing forest, 425 new trees were planted. A few years later, the 217 oldest trees were cut. The forest then contained 1063 trees. How many trees were there before the new trees were planted?

A typical example of the children's written work was the statement 1063 + 217 = 1280 − 425 = 855. This "equation" shows a well-known pattern, that of writing down the operations in the order in which they are carried out and keeping a running-total (Kieran, 1979). To construct an equation that represents the relationships in this problem situation ($\Box$ + 425 − 217 = 1063) requires that the child think precisely the opposite of the way he or she goes about solving the problem. In other words, the problem is solved by adding 217 and subtracting 425, but the equation representing the problem structure involves the addition of 425 and the subtraction of 217.

Solving Open Sentences

Children's first experiences with manipulating strings of arithmetic symbols usually occur within the framework of solving open sentences—sometimes called "missing addend" problems. However, these equations are often presented outside the context of actual word problem situations, with the result that the child lacks a "real-world" support for interpreting them.

Traditionally, in arithmetic, open sentences involving addition and subtraction were given only in their canonical form ($a + b = ?, a - b = ?$). Non-canonical open sentences ($? + b = c, a + ? = c, ? - b = c, a - ? = c$) were delayed until algebra was studied. With the "new math" movement in the early 1960s, such sentences were suddenly introduced in the primary grades. The introduction of non-canonical open sentences in the absence of algebraic methods left open the question of how children might solve such open sentences. Unlike word problem situations where the child's mental representation of the problem could carry with it semantic information on the operation to be carried out, open sentences that are divorced from word problems do not tell the child which operation to perform. For example, children wonder about the operation indicated by the open sentence $3 + ? = 7$. On the surface, it is an addition operation but, from another perspective, it is a subtraction. The child, when presented with such a sentence, can be torn between two different processes.

During the 1970s, researchers investigated children's solving of open sentences by trying to fit a computational model to reaction times. Groen and Poll (1973) presented 30 first graders with addition open sentences in non-canonical form. When the placeholder was in the second position, the children either incremented the given addend or decremented the sum, whichever was quicker. In another study documenting second graders' performance on various canonical and non-canonical forms of addition and subtraction, Nesher (1980) found that non-canonical forms took longer to solve than canonical forms and that the subtractions generally took longer than the additions.

The modeling technique used in these studies has not, except for additions with the unknown in the second position, been able to suggest the actual processes used by young children in solving non-canonical forms of open sentences. The models are based on children's use of counting procedures and do not tell us whether younger elementary schoolchildren are using arithmetic operations (i.e., the given surface operations) or algebraic operations (i.e., the inverses of the given surface operations) to solve open sentences.

Some data on this question do come from studies with older elementary schoolchildren. For example, Booth (1987), in reporting the equation-solving segment of the Children's Mathematical Frameworks project, stated that interviews with six 11-year-olds showed that the majority of them used the given surface operations of the equation and substituted various numbers to find the correct value of the letter. They did not use the inverses of the given operations in either a transposing procedure or a same-operation-to-both-sides procedure.

Complex Arithmetic Expressions

A small number of studies have been carried out with older elementary school children on tasks involving multi-operation arithmetic expressions. Chaiklin and Lesgold (1984) worked with 5 sixth graders who were given the task of judging the equivalence (without computing the total) of three-term arithmetic expressions with a subtraction and an addition operator (e.g., $685 - 492 + 947$, $947 + 492 - 685$, $947 - 685 + 492$, $947 - 492 + 685$). Analyses of thinking-aloud protocols led Chaiklin and Lesgold to suggest that a satisfactory model of pre-algebra students' knowledge of the structure of arithmetic expressions should accommodate the following findings: (a) Students used several different methods of combining numerical terms, even within the same expression, depending on the expression with which it was being compared; (b) they were not able to differentiate between the mathematically correct methods and the other methods they used; (c) they employed both syntactic and semantic methods for judging the equivalence of expressions; and (d) they relied on their knowledge of arithmetic outcomes.

Another part of the above study was carried out by Resnick and a group of researchers in France (Cauzinille-Marmeche, Mathieu, & Resnick, 1984). They presented the same tasks to 11-year-olds just beginning the study of algebra. They found that most students preferred to calculate rather than judge equivalence on the basis of number principles or rules of symbol manipulation. When students did not calculate, their most typical response was to make a judgment about equivalence on the basis of rules for transforming expressions. These rules were often incorrect. The most frequent incorrect rules could be interpreted, according to these researchers, as the dropping or adding of constraints on the principles of commutativity or associativity for addition.

The last study to be included in this section is one that was carried out with older children (12-15 years old) but which included several tasks investigating children's understanding of arithmetic structure and their ability to formalize the procedures they use (Booth, 1984). The major findings, which were based on a large number of individual interviews and teaching experiments with both small groups and whole classes, were the following:

1. Children have difficulty in representing formal mathematical methods even in the arithmetic case. This is partly due to the fact that children often do not make explicit the precise procedures by which they solve problems.

2. The procedures which children use in solving arithmetic problems are often informal methods which are difficult to symbolize concisely.

3. The procedures used are often context-dependent so that they do not readily generalize to other examples (such as algebraic cases), and are symbolized (if at all) in an informal manner which requires reference to the particular context for interpretation.

4. Children consider mathematics to be an empirical subject which requires the production of numerical answers. Even where children can formalize the required procedure and symbolize it correctly, they may not appreciate that this is an appropriate thing to do. (Booth, 1984, pp. 85-86)

We began this introductory section with the statement that "much of school arithmetic is oriented towards 'finding the answer.'" This emphasis in the curriculum allows children to get by with informal, intuitive procedures. However, in algebra, they are required to recognize and use the structure that they have been able to avoid in arithmetic. The ways in which they attempt to cope with the demands of the early part of high school algebra are seen in the remainder of this paper.

ALGEBRA

Children generally begin the study of algebra at about 13 to 15 years of age, depending on the country and programme. The early learning of this mathematical subject usually involves variables, algebraic expressions, equations, and equation solving. As pointed out in the introductory section, children's difficulties with these topics have been found to center on (a) the meaning of letters, (b) the shift to a set of conventions different from those used in arithmetic, and (c) the recognition and use of structure. Studies documenting these difficulties are presented in this section, with special emphasis on those findings related to the recognition and use of structure, for this particular aspect of algebra appears to be one that never really does get sorted out by most students throughout their entire high school algebra career (Wenger, 1987).

The research reviewed herein is sequenced in the order in which topics are generally introduced in the algebra course.

Variables

High school algebra usually starts with instruction in the concept of variable. In elementary school, children have already seen placeholders in open sentences and used letters in formulas such as the area of a rectangle. However, their past experiences cannot easily be related to the many uses of variable to which they are exposed in high school algebra. Usiskin (1988) has described some of these many uses of variable and has related them to the different purposes of algebra.

According to Usiskin, if we consider algebra as generalized arithmetic, then variables can be viewed as pattern generalizers (e.g., in generalizing $3 + 5 = 5 + 3$ to the pattern $a + b = b + a$) and algebraic skills are centered on translating and generalizing known relationships among numbers. If we consider algebra as the study of procedures for solving certain kinds of problems, then variables can be viewed as unknowns (e.g., in translating a word problem into an equation) and algebraic skills involve simplifying and solving. If we consider algebra as the study of relationships between or among quantities, then variables are either arguments or parameters (e.g., in finding an equation for the line through (5,2) with slope 9) and coordinate graphs are often used to represent these relationships. In this conception of algebra, variables truly vary. Finally, if we consider algebra as the study of structures such as groups, rings, integral domains, fields, and vector spaces, then variables are arbitrary objects in a structure related by certain properties.

Thorndike (Thorndike et al., 1923) suggested in the 1920s that different letters be reserved for different interpretations of the variable. However, this suggestion was never seriously considered. Thirty years later, it was still being remarked (Van Engen, 1953) that the symbol x was being used in many different ways, ways which were causing confusion for students. But the new math movement with its emphasis on unifying concepts altered this situation. Ironically, it promoted just the opposite of what Thorndike had earlier suggested. It encouraged the teaching of the concept of variable in its most general form, right from the start. Variables were considered as one of the unifying ideas of the high school algebra curriculum, with all algebraic letters being referred to as variables. Thus, it is not surprising that the same student confusion that was pointed out by Thorndike and by Van Engen should be noted again in the late 1970s. Matz (1979) remarked that lumping together symbolic constants, parameters, unknowns, and unconstrained variables as simply "variables" draws attention only to their common abstract nature. Such an overly general concept of a variable, according to Matz, blurs distinctions that affect how the symbolic value is manipulated, by obscuring restrictions about exactly how and where the variable varies.

The resulting confusion that students experience over the different ways that a single letter variable can be used in algebra often leads to erroneous interpretations. Firth (1975) gave the following task to seventeen 15-year-olds:

If x is any number
  (a) Write the number which is 3 more than x;
  (b) Write the number which is 5 less than x;
  (c) Write the number which is twice as big as x;
  (d) Write the number which is 50% bigger than x.

He found that only 10, 11, 9, and 7 students, respectively, answered the parts correctly. Firth noted that 5 of the 17 students solved the entire task incorrectly by first choosing a value for x and using that value throughout the exercise. It has been suggested that this error may be linked to the student's difficulty in considering $x + 3$ as a final answer. Matz (1979) and Davis (1975) have referred to this difficulty as the process-product dilemma. The operation of adding 3 directly to the letter x can only be expressed in terms of the process: The process is also the product. Another instance of this error is cited in one of Davis's (1975) studies: A student could not operate with a variable because he did not know what number the variable represented. According to Matz and Davis, students must be able to accept the concept of and notation for these suspended operations in order to work with symbolic values. Accepting this concept means changing their expectation about what answers should look like by loosening the arithmetic requirement that an answer is a number.

Collis (1975) found that students, prior to the stage of formal reasoning, work at one of three levels—depending on their experience and mathematical maturity. Children at the lowest level tend to substitute one specific number for a letter (as had 5 of Firth's subjects). If after one trial—in, for example, an equation-solving task—the value does not work, they will give up. At the second level, students are willing to try several numbers in a trial-and-error method. By the time they are at the third level, students have extracted a concept of generalized number by which a symbol x can be regarded as an entity in its own right, but having the same properties as any number with which they have had previous experience.

A large-scale study of some of the various ways in which students use algebraic letters was carried out by Küchemann (1978, 1981) in 1976. As part of the Concepts in Secondary Mathematics and Science (CSMS) project, Küchemann administered a 51-item paper-and-pencil test to 3000 British high school students, aged 13, 14, and 15 years old. He classified each item into one of six levels of interpretation of letters according to the minimum level required for successful performance. Unsuccessful answers were classified into a lower level of interpretation. The six levels that Küchemann used in the analysis of his data, based on levels originally developed by Collis (1975), are the following:

(a) Letter evaluated: The letter is assigned a numerical value from the outset;

(b) Letter not used: The letter is ignored or its existence is acknowledged without giving it a meaning;

(c) Letter used as an object: The letter is regarded as a shorthand for an object or as an object in its own right;

(d) Letter used as a specific unknown: The letter is regarded as a specific but unknown number and can be operated on directly;

(e) Letter used as a generalized number: The letter is seen as representing, or at least being able to take on, several values rather than just one;

(f) Letter used as a variable: The letter is seen as representing a range of unspecified values, and a systematic relationship is seen to exist between two such sets of values.

Küchemann found that, even though the interpretation that students chose to use depended in part on the nature and complexity of the question, most students could not cope consistently with items that required the use of a letter as a specific unknown. They erroneously used one of the three lower-level interpretations instead.

The results of the CSMS algebra research led to a follow-up study, the Strategies and Errors in Secondary Mathematics (SESM) project (Booth, 1984). The aim of this study was to investigate the reasons for the errors that had occurred most frequently in the CSMS study. Interviews with 50 students, as well as teaching experiments with both small groups and entire classes, suggested that—according to Booth—some of the difficulty that students have in interpreting letters as representing generalized numbers may be related to a "cognitive readiness" factor. During the teaching experiments, the lower-ability groups were unable to evolve in their interpretation of letters as the middle- and upper-ability groups did. Another finding of the SESM study was that, even though beginning algebra students were initially unreceptive to the idea of unclosed, non-numerical answers (such as $x + 3$), instruction was quite effective in changing their thinking in this regard.

A different perspective on students' understanding of variables comes from a set of studies that investigated the effects of long-term Logo programming experience on the development of mathematical learning (Hoyles, Sutherland, & Evans, 1985; Noss, 1986; Sutherland, 1987; Sutherland & Hoyles, 1986). As part of the Logo Maths Project, Sutherland investigated the hypothesis that certain programming experiences in Logo provide students with a conceptual basis for variables that enhances their work with traditional paper-and-pencil algebra. The longitudinal case studies of four pairs of pupils (aged 11 to 14) programming in Logo during their normal mathematics lesson, throughout the three years of the project, were followed up by a structured interview at the end of the project. In this interview pupils were asked to (a) make a generalization and formalize it in an algebra context, (b) make a generalization and formalize it in a Logo context, (c) answer algebra questions related to the meaning of letters (taken from the CSMS questionnaire), (d) answer Logo questions related to the meaning of variable names, and (e) represent a function in both Logo and algebra. The results of the paper-and-pencil tests showed that, when compared with a class of non-Logo students and also with 1000 students tested in the CSMS project, the case-study students performed substantially better (Sutherland, 1987).

Algebraic Expressions

After being introduced to the notion of using letters to represent numbers, the next topic in the algebra programme is usually that of operating with these letters in the context of simplifying algebraic expressions. In contrast to the large number of studies carried out on students' concepts of variable, relatively little research has addressed itself specifically to the concept of an algebraic expression.

One study was carried out by Chalouh and Herscovics (1988). In a teaching experiment involving six children individually (12 to 13 years of age), the investigators sought, first, to determine the feasibility of a geometric approach to constructing meaning for algebraic expressions and, second, to uncover the cognitive obstacles associated with such an approach. Based on the work of Collis (1974) and Davis (1975), who both pointed out the incongruencies between arithmetic and algebra and the consequent inability of novice algebra students to regard algebraic expressions as legitimate answers, as well as the research on the difficulties associated with algebraic concatenation (Booth, 1981; Matz, 1979), Chalouh and Herscovics designed an instructional sequence involving algebraic representations for arrays of dots, line segments, and areas of rectangles. It was found that the lessons helped the children to develop meaning for expressions such as $2a + 5a$, but that most of the children were unable to interpret this expression as $7a$. These results suggest that constructing meaning for algebraic expressions does not necessarily lead to spontaneous development of meaning for the simplification of algebraic expressions. This study also showed that beginning algebra students *do* develop an awareness that the conventions used in algebra are different from those used in arithmetic: To one of the questions on the posttest, a child inquired whether or not the researchers wanted her to answer "in algebra."

In contrast to the above study, where the emphasis was on children's construction of meaning for the form of algebraic expressions, other studies have investigated children's structural knowledge of these expressions as evidenced by the processes they use to simplify them. Greeno (1980) has suggested that the process of solving problems involves apprehending the structure of relations in the problem. To test this idea, he carried out a study (Greeno, 1982) with beginning algebra students on tasks involving algebraic expressions. He found that students' performance appeared to be quite haphazard, for a while at least. Their procedures were fraught with unsystematic errors, indicating an absence of knowledge of the structural features of algebra. Their confusion was evident from the way that they partitioned algebraic expressions into component parts. According to Greeno, beginning algebra students are consistent neither in their approach to the testing of conditions before performing some operation nor in their process of performing the operations. For example, they might simplify $4(6x - 3y) + 5x$ as $4(6x - 3y + 5x)$ on one occasion but do something else on another occasion. That a change in the context of a task can lead to a different structuring of the problem was also found in the study carried out by Chaiklin and Lesgold (1984) described earlier.

Algebraic Equations and Equation Solving

Students' difficulties with apprehending the structure of algebraic expressions carry over into their work with the next topic of the programme, algebraic equations. Some of the research on equations and equation solving has dealt explicitly with students' recognition and use of structure; other studies have dealt with structure implicitly. In some studies, knowledge of structure was assumed if the student could recognize the underlying form of an equation (i.e., knowledge of surface structure); in others, knowledge of structure was assumed if the student could discriminate correct equation-solving transformations from incorrect ones (i.e., knowledge of systemic structure). In these latter studies, knowledge of structure included knowledge of (a) the inverse relation between addition and subtraction and between multiplication and division, (b) the equivalence of left-hand and right-hand expressions in an equation, and (c) the equivalence of equations in an equation-solving chain. We shall now look at several studies that have addressed, either implicitly or explicitly, the issue of students' knowledge of the structure of equations and equation solving.

Surface structure. One of the aims of the Algebra Learning Project (Wagner, Rachlin, & Jensen, 1984) was to identify the difficulties that algebra students have in solving standard and nonstandard equations. Two of the tasks were designed to probe students' understanding that the solution to an equation is determined by the surface structure of the equation and not by the symbols used to represent the variable. Drawing on Wagner's (1977, 1981) research on conservation of equation, the investigators asked ninth-grade students to solve the equation $s/8 - 3 = 14$ and then to find the solution to the same equation after an alphabetic transformation of the variable—from s to t. Most students knew immediately that the solution to the equation would not change. In the next task, the literal term t of the equation was changed to $t + 1$, and students were asked for the value of $t + 1$. The majority re-solved the equation, some solving for $t + 1$ directly, but most of them solving for t and then figuring out the value of $t + 1$. In a later task, the students were asked to solve for $(2r + 1)$ in $4(2r + 1) + 7 = 35$. Only one student solved directly for $2r + 1$. The findings of this study show that algebra students have trouble dealing with multiterm expressions as a single unit and suggest that students do not perceive that the basic surface structure of, for example, $4(2r + 1) + 7 = 35$ is the same as for $4x + 7 = 35$.

Many of the studies of equations and equation solving have focused on students' knowledge of parsing (i.e., recognition of the surface structure of an expression or equation). Davis (1975), Davis, Jockusch, and McKnight (1978), Matz (1979), and Greeno (1982), as well as the much earlier studies of Monroe (1915), Rugg and Clark (1918), and Breslich (1939), have all shown that beginning algebra students have enormous difficulties in imposing structure on expressions involving various combinations of operations and numerical and literal terms. The parsing errors seen in the previous section on algebraic expressions—errors such as simplifying $39x - 4$ to $35x$—have been re-documented, as expected, in the studies on novice equation solving. Furthermore, two studies (Carry, Lewis, & Bernard, 1980; Lewis, 1981) have shown the same types of errors with college students. Carry, Lewis, and Bernard, in discussing the error of replacing $2yz - 2y$ by z, have suggested that some students are overgeneralizing certain mathematically valid operations, arriving at a single generic deletion operation that often produces incorrect results. It should be noted that this "deletion" error was the most prevalent error in the Carry, Lewis, and Bernard equation-solving study of college algebra students.

A recent study with a teaching component has shown that instruction can improve students' ability to recognize the form or surface structure of an algebraic equation. Thompson and Thompson (1987) designed a teaching experiment involving two instructional formats: (a) algebraic equation notation and (b) expression trees displayed on a computer screen. After instruction, 8 seventh-grade students did not overgeneralize rules, nor did they fail to perceive the structure of expressions. They also developed a general notion of variable as a placeholder within a structure and the view that the variable could be replaced by anything—a number, another letter, or an expression.

Systemic structure. A teaching experiment carried out by Herscovics and Kieran (1980) emphasized an aspect of the systemic structure of an algebraic equation: the equivalence of left- and right-hand expressions. In a series of individual sessions, 6 seventh- and eighth-grade children were guided in constructing meaning for equations having a variable term on each side of the equation (i.e., for equations with the surface structure $ax \pm b = cx \pm d$). The instructional sequence began by extending the notion of arithmetic equality to include equalities with more than one numerical term on the right side and then hiding selected numbers in these "arithmetic identities." This approach was found to be accessible to the algebra novices of the study and effective in expanding their view of the equal sign from a "do-something signal" (Behr, Erlwanger, & Nichols, 1976) to that of a symbol relating the value on the left side to that on the right (Kieran, 1981).

Another facet of arithmetic/algebraic structure concerns the relationship between operations and their inverses and the equivalent expressions of these relationships. Beginning algebra students are expected to know, for example, that $3 + 4 = 7$ can be expressed as $3 = 7 - 4$ and be able to generalize this knowledge to equations involving literal terms—thereby coming to see, for example, that $x + 4 = 7$ and $x = 7 - 4$ are equivalent and thus have the same solution. That algebra learners have difficulty in judging equivalent equations was shown in a study that included a teaching component with six 12-year-old, beginning algebra students (Kieran, 1984). Two of the errors these students committed—the "Redistribution" error and the "Switching-Addends" error—indicated their confusion with regard to the systemic structure of equations. In the Switching-Addends error, $x + 37 = 150$ was judged to have the same solution as $x = 37 + 150$ (i.e., $x + a = b$ was considered equivalent to $x = a + b$); in the Redistribution error, $x + 37 = 150$ was judged to have the same solution as $x + 37 - 10 = 150 + 10$ (i.e., $x + a = b$ was considered equivalent to $x + a - c = b + c$). In this latter example, the subtraction of 10 on the left side was balanced by the addition of 10 on the right side. Interestingly, the students who committed these two errors were precisely those whose preferred method of equation solving was transposing (i.e., solving $3x + 5 = 23$ by subtracting 5 and then dividing by 3). In contrast, the students who did not commit these errors were those whose preferred method of solution was trial-and-error substitution. That is, those novices who preferred to use the given surface operations in finding the value of the unknown (i.e., the substitution procedure) had a better sense of the relationship between the operations of addition and subtraction than those who preferred to use the inverses of the given surface operations to find the value of the unknown (i.e., the transposing procedure). This finding implies that beginning algebra students who use the transposing procedure to solve equations may be less sure of the underlying structural relationship between addition and subtraction than their use of transposition suggests.

Another aspect of structural knowledge considered to be important in equation solving involves knowledge of equivalence constraints. Greeno

(1982) has pointed out that algebra novices lack knowledge of the constraints that determine whether transformations are permissible. For example, they do not know how to show that an incorrect solution is wrong, except to re-solve the given equation. They do not seem to be aware that an incorrect solution, when substituted into the original equation, will yield different values for the left and right sides of the equation. Nor do they realize that it is only the correct solution that will yield equivalent values for the two expressions in any equation of the equation-solving chain.

However, it is not only novice equation solvers who lack knowledge of these equivalence constraints. Kieran (1984) found that a group of nine experienced, competent high school solvers also lacked this knowledge. However, they had other ways of determining whether two equations have the same solution, without actually solving the equations. They tried to create a surface match of the two equations to be compared, usually by means of transposing terms in one or both of the equations. At the same time that they were trying to create a surface match of the two equations, they were also accessing their mathematical knowledge to find a reason for the equations not to have the same solution. As soon as either process reached its goal, they were able to say whether or not the two equations were equivalent. When the six novices of the study were asked the same question—to determine whether given pairs of equations had the same solution, without actually solving them—they used a different approach. They attempted to pick out what did not match and, on the basis of their arithmetical knowledge, determine whether the mismatches were legal or not. In scanning the equation-pairs for similarities and differences, the experienced algebra students were found to look first at the numerical coefficients and terms, then at the signs, and finally at the positions of the terms. The novices, on the other hand, followed a left-to-right search pattern and rarely seemed able to take in all of the differences between the equations. This inability of beginning algebra students to discriminate the essential features of equations has important consequences for learning theory, as will be seen below.

Equation-solving procedures. Neves (1979) has constructed a model of the learning of equation solving based on the performance of experts. His model, in the form of an adaptive production system, learns from examining worked-out solution procedures. Let us consider the following example:

$$5x + 2 = 3x + 8$$
$$5x = 3x + 8 - 2$$
$$5x - 3x = 6$$
$$2x = 6$$
$$x = 3.$$

The system begins by comparing the first two equations in the chain and notes the differences between them. From this, it infers and constructs a new production: If there is a numerical term, N, on the left-hand side, then subtract N from both sides. The process continues with the subsequent equations.

Simon (1980) has suggested that the logic of this procedure rests on the general principle of applying means-ends analysis. For example:

> We have a certain situation (the first expression of the pair); we want a different situation (the second expression). We detect a difference between the two situations, and we search memory for an operator and examine the resulting expression to see if the difference has been removed. (p. 90)

This emphasis on general skills such as means-ends analysis in the early stages of learning a new domain focuses to a large extent on the characteristics of the problem. But the research evidence provided above shows that this is precisely what the novice is unable to do—that is, the beginning algebra learner is not able to discriminate what is important when first confronted with expressions and equations. Thus, as we have seen, most researchers have tended not to investigate the use of general skills but have looked at how the specific knowledge of novice algebra students influences the nature of the processes they use.

Another large body of research exists where the focus has been on identifying the kinds of procedures used by novices in the solving of equations. Matz (1979) has suggested that two kinds of processes are involved in solving first-degree equations with one unknown: "deductions" and "reductions." The deduction process involves performing the same operation on both sides; the reduction process involves replacing one expression by another equivalent expression. The following example demonstrates both processes:

$$3x + 7 = 2x$$
$$3x + 7 - 2x = 2x - 2x \qquad \text{(deduction)}$$
$$x + 7 = 0. \qquad \text{(reduction)}$$

Swain (1962) has called these two processes "manipulation" and "reduction." The artificial intelligence model of Bundy (1975), used by Carry, Lewis, and Bernard (1980) in their equation-solving study, includes an extra process called "attraction." In the Bundy computer model, $3x + 7 - 2x$ would not be immediately combined into $x + 7$; the two occurrences of the unknown would first be made adjacent, by attraction. The processes identified in these theoretical analyses of equation solving have been investigated in various studies that have looked at students' use of both formal and informal solving procedures.

Whitman (1976) researched the relationship between formal and informal techniques of equation solving with 156 seventh-grade students (six intact classes). She taught them in one of three ways:

(a) Intuitive techniques only: For example, to solve the equation $69 - 96/(7 - a) = 37$, the student would be urged to think:

> 69 minus what number gives 37? — 32
> 96 divided by what number is 32? — 3
> 7 minus what number is 3?   — 4
>
> Solution: $a = 4$;

(b) Formal techniques only: That is, multiply both sides by $(7 - a)$, remove parentheses, combine terms, subtract 259 from both sides, add $69a$ to both sides, and divide both sides by 32;

(c) Intuitive techniques followed by formal techniques.

Whitman found that students who learned to solve equations only intuitively performed better than those who learned both ways in close proximity, while students who learned to solve equations only formally performed worse than those who learned both techniques. Whitman concluded that formal techniques tend to thwart students' intuitive ability to solve equations.

Adi (1978) investigated the relationship of formal and informal techniques of equation solving to Piagetian levels of concrete and formal operations. She used the same two equation-solving methods as Whitman had used. Subjects were 75 prospective elementary school teachers. They were taught first the informal technique (which Adi called the "cover-up" method), then the formal method of equation solving. Subjects in the Piagetian late concrete stage ($n = 26$) and early formal stage ($n = 12$) scored higher than those at the early concrete stage ($n = 37$).

Bell (personal communication, April 1978) has described an intuitive procedure that he has seen students use on equations such as $2x + 9 = 5x$: "Since $2x + 9$ totals $5x$, the 9 must be the same as $3x$ because $2x + 3x$ also equals $5x$. So x is 3."

Petitto (1979) has also observed the use of informal solving processes in her interviews of 7 ninth-grade algebra students. She designated as informal those processes that deal with perceived properties and relationships, and as formal those processes that refer to a linear sequence of steps as described in Matz's work. Petitto found that intuitive techniques often did not generalize—as in equations involving negative numbers. Students who used a combination of formal and informal processes were more successful than those who used only one of these processes.

Equation-solving models. Some studies of equation solving have included different modeling techniques as a method of helping students construct meaning for certain forms of equations and for the operations carried out on those equations. One such study was carried out by O'Brien (1980) who worked with two groups of 23 third-year high school students. They had previously learned some algebra and, at the beginning of the study, were already solving equations using the "Change Side—Change Sign" method. One group was taught meaning for equations and for the manipulations performed on equations by means of concrete materials (bundles of counters and colored cubes). The manipulations involved removing objects from both sides or adding objects to both sides of the concretely modeled equation. The second group was taught meaning for manipulations using a generalization of the addition/subtraction relation (e.g., $2 + 3 = 5$ is equivalent to $2 = 5 - 3$), that is, the Change Side—Change Sign rule. Towards the end of the study, this group also used the rule of doing the same operation to both sides. O'Brien found that the second group became more proficient equation solvers than the concrete-materials group.

Concrete models have also been used by Filloy and Rojano (1985a, 1985b) in teaching experiments aimed at helping students create meaning for equations of the type $Ax \pm B = Cx$ and for the algebraic operations used in solving these equations. Their principal approach was a geometric one, although they also used the balance model in some of their studies. The geometric approach was introduced in the following way:

> A person has a plot of land of dimensions A by x. Next she buys an adjacent plot with an area of B square metres. A second person proposes to exchange this plot for another on the same street having the same area, but a better shape. How much should the depth measure so that the deal is a fair one?

The situation was then represented pictorially (see Figure 1). Teaching interviews with three classes of 12- and 13-year-olds who already knew how to solve equations of the types $x \pm A = B$ and $Ax \pm B = C$ showed that the use of these two concrete models (the balance and the area models) did not significantly increase most students' ability to operate at the symbolic level with equations having two occurrences of the unknown. The well-known equation-solving error of combining constants and coefficients was also seen in this study, in particular with the use of the geometric model. Students tended to fixate on the model and seemed unable to apply previous

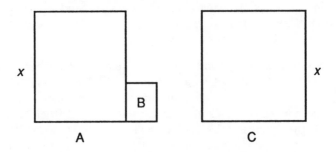

Figure 1. Pictorial representation of a geometric situation used to model equations of the form $Ax + B = Cx$.

equation-solving knowledge to the simplified equations of the instructional sequence.

A final study to be discussed in this section on equation solving is one which did not use concrete or pictorial models but, rather, drew on the numerical approach used in an earlier teaching experiment by Herscovics and Kieran (1980). At the outset of the study, Kieran (1988) pretested six average-ability seventh-graders who had not had any previous algebra instruction. She found that the students showed two different equation-solving preferences, both based on their elementary school experience with open sentences. Some preferred to solve the simple equations of the pretest by means of arithmetic methods such as substitution and known number facts; others preferred transposing (and in fact seemed unaware of the potential of substitution as an equation-solving procedure). Those who preferred substitution viewed the letter in an equation as representing a number in a balanced equality relationship; those who preferred transposing viewed the letter as having no meaning until its value was found by means of certain inverse operations (see Kieran, 1983, for more details). In the teaching experiment on equation solving that followed, the procedure of performing the same operation on both sides of an equation was carried out first on arithmetic equalities (e.g., $10 + 7 = 17$), and then on the algebraic equations built from these arithmetic equalities (e.g., $x + 7 = 17$):

$$10 + 7 = 17 \qquad\qquad x + 7 = 17$$
$$\downarrow \qquad\qquad\qquad \downarrow$$
$$10 + 7 - 7 = 17 - 7 \qquad\qquad x + 7 - 7 = 17 - 7.$$

Kieran found that those students who had initially preferred transposing were in general unable to make sense of the solving procedure being taught, that is, the procedure of performing the same operation on both sides of an algebraic equation. This suggests that, although transposing is considered by many teachers to be a shortened version of the procedure of performing the same operation on both sides, these two solving methods may be perceived quite differently by beginning algebra students. The procedure of performing the same operation on both sides of an equation emphasizes the symmetry of an equation; this emphasis is quite absent in the procedure of transposing.

Although the above investigation involved only six case studies of beginning algebra learners, the findings suggest that there may not be just one path that is followed in the learning of algebra. Some learners focus initially on the given surface operations and on the relationship of equality between left- and right-hand expressions of an equation; these students may be more open to the solving procedure of performing the same operation on both sides. Other learners focus immediately on transposing and on the inverses of the given surface operations; they may prefer to solve equations, not by the same-operation-to-both-sides method, but by extending their transposing method.

An interesting implication of these findings is in the field of algebra word problem research. I hypothesize that learners whose focus is on transposing and who actually think in terms of the solving operations when tackling arithmetic word problems are precisely those who might experience the greatest difficulties when attempting to represent algebra word problems by means of equations. The solver who is predisposed towards transposing may experience a conflict between the operations to be used in generating an equation that represents the problem structure and the operations to be used in solving the problem. This difficulty was hinted at earlier in our discussion of the problem-solving work of sixth-graders, reported by Vergnaud and his colleagues. (For a systematic review of the research on algebra word problems, see the paper by Chaiklin in this volume.)

CONCLUDING REMARKS

The early learning of algebra involves grappling with the topics of variables, algebraic expressions, equations, and equation solving. The research discussed in this paper has shown that students' difficulties with these topics center on the meaning of letters, the shift to a set of conventions different from those used in arithmetic, and the recognition and use of structure. It has been found that some of these difficulties are amenable to instruction; others less so. One particularly troublesome area concerns the understanding of certain structural aspects of school algebra. For example, the equality relationship between left- and right-hand expressions of equations is a cornerstone of much of the algebra instruction currently taking place. It is the basis of many of the concrete models used to represent equations and equation solving; it is also an integral part of the symmetric procedure of performing the same operation on both sides of the equation. However, it has been found that, for some students, teaching methods based on this aspect of the structure of equations often do not succeed. For these students—who tend to view the right side of an equation as the answer and who prefer to solve equations by transposing—the equation is simply not seen as a balance between right and left sides nor as a structure that is operated on symmetrically. That understanding seems to be absent. These same students also appear to have difficulty in formalizing even such simple relationships as the equivalent forms of addition and subtraction. Another often-documented finding concerns the inability of beginning algebra students to "see" the surface structure of algebraic expressions containing various combinations of operations and literal terms. This difficulty seems to continue throughout the algebra career of many students, as evidenced by errors such as reducing $(a + b + c)/(a + b)$ to c, seen among college students. In conclusion, many high school algebra students appear to be experiencing serious obstacles in their ability to recognize and use the structure of algebra.

Most of the studies reviewed in this chapter were carried out in traditional, non-computer settings. One might then be inclined to ask whether these findings would be relevant in a modified algebra environment that relied heavily on the use of the computer as a problem-solving tool. In this modified environment, there would likely be considerably less emphasis on manipulation of algebraic expressions and solving of equations. Nevertheless, I believe that the results of much of the research discussed in this chapter would be applicable to computer-intensive algebra learning situations for the following reason: In this hypothetical algebra programme, there would probably still be the need to represent formal mathematical methods, that is, to formalize procedures and to symbolize them. Indeed, Fey (1984) points out:

> As procedural operations are increasingly mechanized [by computers], there remains an important task of conceptualization and planning. Problems must still be identified and cast in mathematical form; the proper analyses must be structured and the results of computer-assisted calculations must be properly tested and interpreted. To perform this fundamental role, individuals must have a sound understanding of the scope and structure of available mathematical methods. (p. 28)

The challenge to researchers is to devise studies that will push forward our knowledge of how students can come to understand the structure of elementary algebra and algebraic methods.

REFERENCES

Adi, H. (1978). Intellectual development and reversibility of thought in equation solving. *Journal for Research in Mathematics Education, 9*, 204-213.

Behr, M., Erlwanger, S., & Nichols, E. (1976). *How children view equality sentences* (PMDC Technical Report No. 3). Tallahassee: Florida State University. (ERIC Document Reproduction Service No. ED144802)

Booth, L. R. (1981). Child-methods in secondary mathematics. *Educational Studies in Mathematics, 12*, 29-41.

Booth, L. R. (1984). *Algebra: Children's strategies and errors.* Windsor, Berkshire: NFER-Nelson.

Booth, L. R. (1987). Equations revisited. In J. C. Bergeron, N. Herscovics, & C. Kieran (Eds.), *Proceedings of the Eleventh International Conference for the Psychology of Mathematics Education* (Vol. I, pp. 282-288). Montréal, Québec, Canada: Université de Montréal.

Breslich, E. R. (1939). Algebra, a system of abstract processes. In C. H. Judd (Ed.), *Education as cultivation of the higher mental processes.* New York: Macmillan.

Briars, D. J., & Larkin, J. H. (1984). An integrated model of skill in solving elementary word problems. *Cognition and Instruction, 1*, 245-296.

Bundy, A. (1975). *Analysing mathematical proofs* (DAI Research Report No. 2). Edinburgh, Scotland: University of Edinburgh, Department of Artificial Intelligence.

Carpenter, T. P., & Moser, J. M. (1982). The development of addition and subtraction problem-solving skills. In T. P. Carpenter, J. M. Moser, & T. A. Romberg (Eds.), *Addition and subtraction: A cognitive perspective* (pp. 9-24). Hillsdale, NJ: Lawrence Erlbaum.

Carry, L. R., Lewis, C., & Bernard, J. (1980). *Psychology of equation solving: An information processing study* (Final Technical Report). Austin: University of Texas at Austin, Department of Curriculum and Instruction.

Cauzinille-Marmeche, E., Mathieu, J., & Resnick, L. B. (1984, April). *Children's understanding of algebraic and arithmetic expressions.* Paper presented at the annual meeting of the American Educational Research Association, New Orleans, LA.

Chaiklin, S., & Lesgold, S. (1984, April). *Prealgebra students' knowledge of algebraic tasks with arithmetic expressions.* Paper presented at the annual meeting of the American Educational Research Asssociation, New Orleans, LA.

Chalouh, L., & Herscovics, N. (1988). Teaching algebraic expressions in a meaningful way. In A. Coxford (Ed.), *The ideas of algebra, K-12* (1988 Yearbook, pp. 33-42). Reston, VA: National Council of Teachers of Mathematics.

Collis, K. F. (1974, June). *Cognitive development and mathematics learning.* Paper presented at the Psychology of Mathematics Workshop, Centre for Science Education, Chelsea College, London.

Collis, K. F. (1975). *The development of formal reasoning.* Newcastle, Australia: University of Newcastle.

Davis, R. B. (1975). Cognitive processes involved in solving simple algebraic equations. *Journal of Children's Mathematical Behavior, 1*(3), 7-35.

Davis. R. B., Jockusch, E., & McKnight, C. (1978). Cognitive processes in learning algebra. *Journal of Children's Mathematical Behavior, 2*(1), 10-320.

Fey, J. T. (Ed.). (1984). *Computing and mathematics: The impact on secondary school curricula.* College Park: The University of Maryland.

Filloy, E., & Rojano, T. (1985a). Obstructions to the acquisition of elemental algebraic concepts and teaching strategies. In L. Streefland (Ed.), *Proceedings of the Ninth International Conference for the Psychology of Mathematics Education* (pp. 154-158). Utrecht, The Netherlands: State University of Utrecht.

Filloy, E., & Rojano, T. (1985b). Operating on the unknown and models of teaching. In S. K. Damarin & M. Shelton (Eds.), *Proceedings of the Seventh Annual Meeting of PME-NA* (pp. 75-79). Columbus: Ohio State University.

Firth, D. E. (1975). *A study of rule dependence in elementary algebra.* Unpublished master's thesis, University of Nottingham, England.

Greeno, J. G. (1980). Trends in the theory of knowledge for problem solving. In D. T. Tuma & F. Reif (Eds.), *Problem solving and education: Issues in teaching and research* (pp. 9-23). Hillsdale, NJ: Lawrence Erlbaum.

Greeno, J. G. (1982, March). *A cognitive learning analysis of algebra.* Paper presented at the annual meeting of the American Educational Research Association, Boston, MA.

Groen, G. J., & Poll, M. (1973). Subtraction and the solution of open sentence problems. *Journal of Experimental Child Psychology, 16,* 292-302.

Herscovics, N., & Kieran, C. (1980). Constructing meaning for the concept of equation. *Mathematics Teacher, 73,* 572-580.

Hoyles, C., Sutherland, R., & Evans, J. (1985). *The Logo Maths Project: A preliminary investigation of the pupil-centred approach to the learning of Logo in the secondary mathematics classroom, 1983-4.* London: University of London, Institute of Education.

Kieran, C. (1979). Children's operational thinking within the context of bracketing and the order of operations. In D. Tall (Ed.), *Proceedings of the Third International Conference for the Psychology of Mathematics Education.* Coventry, England: Mathematics Education Research Centre, Warwick University.

Kieran, C. (1981). Concepts associated with the equality symbol. *Educational Studies in Mathematics, 12,* 317-326.

Kieran, C. (1983). Relationships between novices' views of algebraic letters and their use of symmetric and asymmetric equation-solving procedures. In J. C. Bergeron & N. Herscovics (Eds.), *Proceedings of the Fifth Annual Meeting of PME-NA* (Vol. 1, pp. 161-168). Montréal, Québec, Canada: Université de Montréal.

Kieran, C. (1984). A comparison between novice and more-expert algebra students on tasks dealing with the equivalence of equations. In J. M. Moser (Ed.), *Proceedings of the Sixth Annual Meeting of PME-NA* (pp. 83-91). Madison: University of Wisconsin.

Kieran, C. (1988). Two different approaches among algebra learners. In A. Coxford (Ed.), *The ideas of algebra, K-12* (1988 Yearbook, pp. 91-96). Reston, VA: National Council of Teachers of Mathematics.

Kintsch, W., & Greeno, J. G. (1985). Understanding and solving word arithmetic problems. *Psychological Review, 92*(1), 109-129.

Küchemann, D. (1978). Children's understanding of numerical variables. *Mathematics in School, 7*(4), 23-26.

Küchemann, D. (1981). Algebra. In K. Hart (Ed.), *Children's understanding of mathematics: 11-16* (pp. 102-119). London: John Murray.

Lewis, C. (1981). Skill in algebra. In J. R. Anderson (Ed.), *Cognitive skills and their acquisition*. Hillsdale, NJ: Lawrence Erlbaum.

Lindvall, C. M., & Ibarra, C. G. (1978, March). *An analysis of incorrect procedures used by primary grade pupils in solving open addition and subtraction sentences.* Paper presented at the annual meeting of the American Educational Research Association, Toronto, Ontario. (ERIC Document Reproduction Service No. ED155049)

Matz, M. (1979). *Towards a process model for high school algebra errors* (Working Paper 181). Cambridge: Massachusetts Institute of Technology, Artificial Intelligence Laboratory.

Mayer, R. E. (1980). *Schemas for algebra story problems* (Report No. 80-3). Santa Barbara: University of California, Department of Psychology, Series in Learning and Cognition.

Monroe, W. S. (1915). A test of the attainment of first-year high-school students in algebra. *School Review, 23*, 159-171.

Nesher, P. (1980). The internal representation of open sentences in arithmetic. In R. Karplus (Ed.), *Proceedings of the Fourth International Conference for the Psychology of Mathematics Education* (pp. 271-278). Berkeley: University of California.

Nesher, P., Greeno, J. G., & Riley, M. S. (1982). The development of semantic categories for addition and subtraction. *Experimental Studies in Mathematics, 13*, 373-394.

Neves, D. (1979, May). *Learning algebra from a textbook.* Paper presented at the Conference on Cognitive Processes in Algebra, University of Pittsburgh.

Noss, R. (1986). Constructing a conceptual framework for elementary algebra through Logo programming. *Educational Studies in Mathematics, 17*, 335-357.

O'Brien, D. J. (1980). *Solving equations.* Unpublished master's thesis, University of Nottingham, England.

Petitto, A. (1979). The role of formal and non-formal thinking in doing algebra. *Journal of Children's Mathematical Behavior, 2*(2), 69-82.

Riley, M. S., & Greeno, J. G. (1978, May). *Importance of semantic structure in the difficulty of arithmetic word problems.* Paper presented at the annual meeting of the Midwestern Psychological Association, Chicago, IL.

Rugg, H. O., & Clark, J. R. (1918). Scientific method in the reconstruction of ninth-grade mathematics. *Supplementary Educational Monographs, 2*(1, Whole No. 7).

Simon, H. A. (1980). Problem solving and education. In D. T. Tuma & F. Reif (Eds.), *Problem solving and education: Issues in teaching and research* (pp. 81-96). Hillsdale, NJ: Lawrence Erlbaum.

Sutherland, R. (1987). A study of the use and understanding of algebra related concepts within a Logo environment. In J. C. Bergeron, N. Herscovics, & C. Kieran (Eds.), *Proceedings of the Eleventh International Conference for the Psychology of Mathematics Education* (Vol. I, pp. 241-247). Montréal, Québec, Canada: Université de Montréal.

Sutherland, R., & Hoyles, C. (1986). Logo as a context for learning about variable. *Proceedings of the Tenth International Conference of the Psychology of Mathematics Education* (pp. 301-306). London: University of London, Institute of Education.

Swain, R. L. (1962). The equation. *Mathematics Teacher, 55*, 226-236.

Thompson, P., & Thompson, A. (1987). Computer presentations of structure in algebra. In J. C. Bergeron, N. Herscovics, & C. Kieran (Eds.), *Proceedings of the Eleventh International Conference for the Psychology of Mathematics Education* (Vol. I, pp. 248-254). Montréal, Québec, Canada: Université de Montréal.

Thorndike, E. L., Cobb, M. V., Orleans, J. S., Symonds, P. M., Wald, E., & Woodyard, E. (1923). *The psychology of algebra.* New York: Macmillan.

Usiskin, Z. (1988). Conceptions of school algebra and uses of variables. In A. Coxford (Ed.), *The ideas of algebra, K-12* (1988 Yearbook, pp. 8-19). Reston, VA: National Council of Teachers of Mathematics.

Van Engen, H. (1953). The formation of concepts. In H. F. Fehr (Ed.), *The learning of mathematics: Its theory and practice* (Twenty-first Yearbook). Washington, DC: National Council of Teachers of Mathematics.

Vergnaud, G. (1982). A classification of cognitive tasks and operations of thought involved in addition and subtraction problems. In T. P. Carpenter, J. M. Moser, & T. A. Romberg (Eds.), *Addition and subtraction: A cognitive perspective* (pp. 39-59). Hillsdale, NJ: Lawrence Erlbaum.

Vergnaud, G., Benhadj, J., & Dussouet, A. (1979). *La coordination de l'enseignement des mathématiques entre le cours moyen 2e année et la classe de sixième.* Paris: Institut National de Recherche Pédagogique.

Wagner, S. (1977, April). *Conservation of equation and function and its relationship to formal operational thought.* Paper presented at the annual meeting of the American Educational Research Association, New York.

Wagner, S. (1981). Conservation of equation and function under transformations of variable. *Journal for Research in Mathematics Education, 12,* 107-118.

Wagner, S., Rachlin, S. L., & Jensen, R. J. (1984). *Algebra Learning Project: Final report.* Athens: University of Georgia, Department of Mathematics Education.

Wenger, R. H. (1987). Cognitive science and algebra learning. In A. Schoenfeld (Ed.), *Cognitive science and mathematics education* (pp. 115-135). Hillsdale, NJ: Lawrence Erlbaum.

Whitman, B. S. (1976). Intuitive equation solving skills and the effects on them of formal techniques of equation solving (Doctoral dissertation, Florida State University, 1975). *Dissertation Abstracts International, 36,* 5180A. (University Microfilms No. 76-2720)

ACKNOWLEDGMENT

The author wishes to thank Lesley Booth for her reaction to an earlier version of this paper.

A Question of Structure
or
A Reaction to: "The Early Learning of Algebra: A Structural Perspective"

Lesley R. Booth

Department of Pedagogics and Scientific Studies in Education
James Cook University of North Queensland

The identification in Kieran's paper of students' difficulties in algebra as relating primarily to (a) the meaning of letters, (b) changes from arithmetical conventions, and (c) the recognition and use of structure, reaffirms the need to distinguish between the syntactic and semantic aspects of algebra (see Kaput, this volume; Resnick, Cauzinille-Marmeche, & Mathieu, 1987).

One of the main functions of algebra is the representation of general relations and procedures in concise and unambiguous terms. The value of this kind of representation is twofold. In the first place, it enables us to apply these relations and procedures in a wide range of problems to which they are relevant. In the second place, it enables us to rearrange or recombine relations and other algebraic entities, thereby deriving new relations that may be useful in solving new kinds of problems or helpful in furthering our understanding of the original relations. Symbolic representation fulfills both functions by objectifying or making explicit the mathematical relations involved, so that these can become the objects of deliberate scrutiny, reflection, and manipulation. In this way we can (a) examine the structural similarities between relations and problem situations to determine the relevance of one to the other, and (b) manipulate the algebraic elements into new configurations. The role of symbolic representation thus assumes considerable importance. Indeed, the most obvious feature of algebra is its use of letters, its introduction of new notation and convention, and its focus upon the manipulation of terms and the simplification of expressions—in other words, its *syntax*.

While the ability to manipulate symbols is important in algebra, it is clear that the critical aspect of such work is understanding what the algebraic statements represent and what rationale and justification underlie the transformations allowed. For example, before we can apply a given relation to a particular problem situation, we must first apprehend a structural isomorphism between the problem situation and the relation. How else do we know that a particular relation is the one required, and not some other? Similarly, our ability to manipulate algebraic symbols successfully requires that we first understand the structural properties of mathematical operations and relations which distinguish allowable transformations from those that are

57

not. These structural properties constitute the *semantic* aspects of algebra. Without an understanding of the semantics of algebra, the mere manipulation of symbols becomes a fairly arbitrary exercise in symbol gymnastics, sometimes performed correctly and sometimes not, but in either case with little sense of purpose. The essential feature of algebraic representation and symbol manipulation, then, is that it should *proceed from* an understanding of the semantics or referential meanings that underlie it.

However, substantial evidence exists to indicate that the learning of algebra is addressed by many students as a problem of learning to manipulate symbols in accordance with certain transformation rules (i.e., syntactically) without reference to the meaning of the expressions or transformations (i.e., the semantics). This is, of course, not surprising, since most algebra syllabuses in the past have paid considerable attention to the syntactic aspects of algebra, precisely because of the central role that symbolic representation plays in algebraic work, because power over such representation is crucial to successful performance in algebra, and because the symbolism is both new to the students and an obvious feature of this area of study. The mathematical relations which are the real object of algebraic representation, and which provide the justification for the manipulations and simplifications involved, are implicitly assumed to be familiar to the students from their work in arithmetic and so are given little attention. The obvious difficulties that students face in learning algebra have naturally been attributed to what is the focus of study, namely the introduction and manipulation of the symbols themselves. Students' difficulties in algebra, it has generally been assumed, are largely difficulties in learning the syntax.

Over the past decade, however, research evidence has been accumulating to indicate that many students have a very poor understanding of the relations and mathematical structures that are the basis of algebraic representation. This lack of understanding is not a new "algebraic" phenomenon: The research summarised by Kieran shows that the problem has its origins in arithmetic. Indeed, a major part of students' difficulties in algebra stems precisely from their lack of understanding of arithmetical relations. The ability to work meaningfully in algebra, and thereby handle the notational conventions with ease, requires that students first develop a semantic understanding of arithmetic.

One task for research is to examine the whole question of students' recognition and use of structure and how this recognition may develop. A second task is to use this information to devise new learning activities and environments to assist students in this development. A start on the first task has been made by, among others, a research study currently being conducted at James Cook University (Booth, 1987). It is to be hoped that the studies being conducted elsewhere on the application of computer environments to the learning of algebra will provide us with significant insights into the second task.

REFERENCES

Booth, L. R. (1987). *Grade 8/9 students' understanding of structural properties in mathematics* (Special Research Grant). Townsville, Queensland, Australia: James Cook University of North Queensland. (Project currently being extended to Grades 6/7)

Resnick, L. B., Cauzinille-Marmeche, E., & Mathieu, J. (1987). Understanding algebra. In J. Sloboda & D. Rogers (Eds.), *Cognitive processes in mathematics*. Keele, England: Oxford University Press.

Cognitive Obstacles Encountered in the Learning of Algebra

Nicolas Herscovics
Department of Mathematics
Concordia University, Montreal

THE NOTION OF A COGNITIVE OBSTACLE

Assessment studies in both Great Britain (Hart, 1981) and the United States (Carpenter, Corbitt, Kepner, Lindquist, & Reys, 1981) indicate that algebra is a major stumbling block for many students in secondary school. Only a minority of pupils completing an introductory course achieve a reasonable grasp of the course content. Even fewer manage to build up enough courage for a second course. The extent of the problem warrants a careful examination of the possible causes.

Two types of arguments are often used to explain the extensive failure rate in high school algebra. Some people are convinced that it is due essentially to inadequate teaching. They strongly believe that the problem is mainly one of instruction and that "if only we could train teachers to teach well, most students would understand." This viewpoint is both optimistic and simplistic, for it presumes that all the problems involved in the learning of algebra can be solved and that these are primarily problems of communication. A second type of argument shifts the onus from teacher to student. The explanation given is that "mathematics, including algebra, was never intended for the general population and, thus, we need not even bother teaching all students." This perspective is rather elitist for it suggests limiting algebra to the few who can cope with it in its traditional presentation. However, it is difficult to accept such views in a modern society, for it is nearly impossible to enter a great many professions without the algebraic tools necessary to mathematize most of the concepts involved in a given field, that is, to translate them into mathematical models.

There is yet another possible explanation for the difficulties encountered in the learning of algebra—one that denigrates neither the teacher's professionalism nor the student's intellectual potential. It is an explanation that takes into account historical, epistemological, and psychological points of view. In his *History of Mathematics*, Boyer (1968/1985) has brought out the fact that high school algebra is a relatively young discipline in comparison with arithmetic and geometry. Classical algebra was introduced about the year 830 A.D. by Al-Khowarizmi in the Middle East. It was presented as a list of rules and procedures needed to solve specific linear and quadratic equations. Modern notation and an analytic approach were ushered in about

60

750 years later by the Frenchman, François Viète, in 1579. Thus, historically, the birth of algebra was late and its growth was rather slow. This lengthy period of development can be attributed to its relatively more abstract nature, its introduction of new concepts that were no longer simply numerical or figural, and its increasingly symbolic representation.

Another perspective for examining the difficulties encountered in the learning of algebra is an epistemological one. The word *epistemology* is used here in the sense of a historical and critical study of science, in particular, and the growth of knowledge, in general. In his book on the structure of scientific revolutions, Kuhn (1970) contends that science is not just a steady, incremental accumulation of knowledge but, rather, a history of violent, intellectual revolutions in which one conceptual view is replaced by another. To grasp the meaning of *scientific revolution*, one need only look at the massive impact of a few scientists—Galileo and his influence on astronomy, Newton and his contribution to classical mechanics, and more recently, Einstein and his theory of relativity.

The French philosopher Bachelard (1938/1983) has studied the conditions leading to the development of scientific thinking. He has introduced the notion of *epistemological obstacle*:

> When one looks for the psychological conditions of scientific progress, one is soon convinced that *it is in terms of obstacles that the problem of scientific knowledge must be raised* [italics added]. The question here is not that of considering external obstacles, such as the complexity and transience of phenomena, or to incriminate the weakness of the senses and of the human spirit; it is in the very act of knowing, intimately, that sluggishness and confusion occur by a kind of functional necessity. It is there that we will point out causes of stagnation and even regression; it is there that we will reveal causes of inertia which we will call *epistemological obstacles* [italics added]. (p. 13, translation by the author of this paper)

In his many books on the subject, Bachelard has identified several kinds of epistemological obstacles: the tendency to rely on deceptive intuitive experiences; the tendency to generalize, which may hide the particularity of an event; the obstacles caused by natural language; and many more. He has thoroughly substantiated each type of obstacle with examples from the history of science.

Bachelard has defined epistemological obstacles in the context of the development of scientific thinking in general, not in terms of specific, individual learning experiences. However, just as the development of science is strewn with *epistemological obstacles*, the acquisition of new conceptual schemata by the learner is strewn with *cognitive obstacles*. And, just as epistemological obstacles are considered normal and inherent to the development of science, so should cognitive obstacles be considered normal and inherent to the learner's construction of knowledge. The word *cognitive* is used here to distinguish it from the word *epistemological*, a term that has acquired a distinct meaning from the historical context.

The notion of cognitive obstacle is a promising one, with many pedagogical implications. However, in order to be construed as a natural occurrence, it needs to be related to a learning theory that takes into account the initial construction of new knowledge. Piaget's theory of equilibration would seem to provide a suitable framework. From the Piagetian perspective, the acquisition of knowledge is a process involving a constant interaction between the learning subject and his or her environment. This process of equilibration involves not only *assimilation*—the integration of the things to be known into some existing cognitive structure—but also *accommodation*—changes in the learner's cognitive structure necessitated by the acquisition of new knowledge. However, the learner's existing cognitive structures are difficult to change significantly, their very existence becoming cognitive obstacles in the construction of new structures.

These theoretical considerations can be illustrated by examining the difficulties many students experience at the beginning of their initial algebra course. Several researchers have found that pupils' prior arithmetical experience can be a source of difficulty in their construction of meaning for algebraic expressions. Collis (1974) has pointed out that beginning algebra students view algebraic expressions as statements that are somehow incomplete. He has explained this perception in terms of the students' inability to hold unevaluated operations in suspension. For instance, younger children require that two numbers connected by an operation be actually replaced by the result of that operation. Later, when they are introduced to algebra, expressions such as $x + 7$ and $3x$ cannot be replaced by a third number and, in that sense, cannot be "closed." By the age of 15, students can hold unevaluated operations in suspension. Collis refers to this as their "acceptance of lack of closure."

Davis (1975) has also brought out some incongruencies between arithmetic and algebra that make it difficult for students to regard algebraic expressions as "legitimate" answers. This difficulty is related to the distinction between arithmetic addition and algebraic addition. In *arithmetic addition*, $3 + 5$ is viewed as the problem or question, and 8 is viewed as the answer. However, in *algebraic addition*, as in $x + 7$, the expression describes both the operation of adding 7 to x, as well as the result that will be obtained when one carries out the operation. Thus, to the student with an arithmetic frame of reference, accepting algebraic expressions as answers requires a major cognitive adjustment, that of overcoming what Davis calls the "name-process" dilemma—wherein the expression both describes the *process* and *names* the answer.

Concatenation, that is, the juxtaposition of two symbols, is another source of difficulty for the beginning algebra student. Matz (1979) has observed that, in arithmetic, concatenation denotes implicit addition, as in both place-value numeration and mixed-numeral notation. However, in algebra, concatenation denotes multiplication. This explains why—one month after a careful introduction to concatenation—several students, when asked to substitute 2 for *a* in 3*a*, thought that the result would be 32. Only when specifically requested to respond "in algebra" did they reply "3 times 2" (Chalouh & Herscovics, 1988). The tendency of pupils to fall back into an arithmetic frame of reference is a strong reminder of the words of Kuhn and Bachelard. It illustrates how difficult it is, not just in the history of science but also in the process of individual learning, to overcome existing conceptual frameworks in order to construct new ones and how, even then, the old and the new frames of reference conflict with each other.

Because another paper (Kieran, this volume) reviews the research literature on the early learning of algebra, this paper deals with the learning of higher-order algebraic concepts, namely, equations in two variables, their graphs, and the notion of function. The introduction of these concepts is not a simple generalization of arithmetic nor a simple extension of single-variable algebra. Some new ideas are involved which cannot easily be assimilated into the learner's existing cognition. New cognitive structures must be constructed, and this process of accommodation may necessarily confront the student with major cognitive obstacles. The identification of these obstacles is our prime objective.

TRANSLATION FROM NATURAL LANGUAGE INTO EQUATIONS

Evidence of the Existence of a Cognitive Problem

In the last few years, the teaching of mathematics through problem solving has been greatly emphasized. In a problem-solving context, the generation of algebraic equations can be viewed as a process of mathematical modeling. However, as will be seen, students encounter major difficulties in representing word problems by equations. It is tempting to believe that after extensive work with equations in one unknown, students will have an easier time constructing expressions and equations in two variables. But, the presence of more than one variable seems to compound their difficulties. This finding is revealed in two large-scale assessment studies.

Küchemann (1978, 1981) has reported the results of algebra tests involving 3000 British students in their second, third, and fourth years of secondary school (mean ages of 13.3, 14.3, and 15.3 years, respectively). British pupils start secondary school at age 12 and begin learning algebra quite early, at about age 12 or 13. Thus, by the age of 15, they are often taking a second mathematics course that includes algebra. In the study reported by Küchemann, all students were asked the following question:

> Blue pencils cost 5 pence each and red pencils cost 6 pence each. I buy some blue and some red pencils and altogether it costs me 90 pence. If *b* is the number of blue pencils bought, and if *r* is the number of red pencils bought, what can you write down about *b* and *r*?

The percentages of correct responses within each age group were 2%, 11%, and 13%, respectively. These results suggest that, when expressing a relatively simple relationship between two variables, students in fourth year had as many difficulties as those in third year. The most common wrong answer (17% of all students) was $b + r = 90$. This kind of mistake indicates that many students have a tendency to view literal symbols as representing sets, with b representing the set of blue pencils and r the set of red ones. The letters are used here merely as labels identifying specific sets.

A similar but even simpler problem was used in the 1977-78 National Assessment of Educational Progress (NAEP), which tested over 70,000 American pupils (9, 13, and 17 years of age). The problem was:

> Carol earned D dollars during the week. She spent C dollars for clothes and F dollars for food. Write an expression using D, C and F that shows the number of dollars she has left.

This question should be almost trivial for 17-year-old students who have had one or two years of algebra. It involves a simple syntactic translation of the given information into an algebraic expression. Here the literal symbols represent specific quantities that need not be related numerically by an equation, as required in the previous problem. Only two straightforward operations are involved. Yet, a third of the 17-year-olds who had had one year of algebra and over a quarter of those who had had two years of algebra could not provide an acceptable answer (Carpenter et al., 1981).

In addition to modeling real-life situations, the generation of appropriate algebraic equations can also be viewed from another perspective, that of translating from one symbol system (in this case, natural language) to another (algebra). For beginning algebra students, the problem of translation is amplified by the fact that they are not yet familiar with the new symbol system and its specific semantic structure. The cognitive problems involved in this translation process are not restricted to novices. In fact, they persist into later years, even among scientifically oriented college students.

The Student-Professor Problem

The persistence of translation problems has been brought to light by Clement and his colleagues in a series of papers (Clement, 1982; Clement, Lochhead, & Monk, 1981; Clement, Lochhead, & Soloway, 1979) reporting results of a test given to 150 freshman engineering students at a major state university. They were asked the following two questions:

- Write an equation using the variables S and P to represent the following statement: "There are six times as many students as professors at this university." Use S for the number of students and P for the number of professors.
- Write an equation using the variables C and S to represent the

following statement: "At Mindy's restaurant, for every four people who order cheesecake, there are five people who order strudel." Let C represent the number of cheesecakes and S represent the number of strudels ordered.

The rate of correct responses was 63% for the first problem and 27% for the second one. In both cases, 68% of the errors were reversals: $6S = P$ instead of $S = 6P$ and $4C = 5S$ instead of $5C = 4S$. The investigators demonstrated that these were not careless mistakes, that the students understood the problems, and that they possessed adequate algebraic skills. The high incidence of the reversal error was quite surprising and the authors investigated whether it might occur in the reverse translation from algebra to English. Clement, Lochhead, and Monk (1981) asked 83 engineering students to write a sentence in English that gives the same information as the equation $A = 7S$, where A is the number of assemblers in a factory and S is the number of solderers. Of the given sample, 71% wrote sentences that indicated there were more solderers than assemblers, hence showing, once again, the reversal error. Lochhead (1980) gave this problem to 202 university faculty members and 148 high school teachers and found that 35% and 47%, respectively, gave incorrect interpretations. That the percentage of reversal errors is even higher when interpreting an equation than when generating one was also found in a cross-cultural study (Mestre & Lochhead, 1983).

Identification of Possible Cognitive Obstacles

Clement and his colleagues have also investigated the types of thinking that lead to such mistakes. By analyzing the transcripts of videotaped interviews of individual students, they have identified two basic sources: a syntactic type of thinking and a semantic type. In a *syntactic* translation, the student assumes that the sequence of words maps directly into a corresponding sequence of literal symbols in an equation (see Figure 1). In a *semantic* translation, students link the equation to the meaning of the problem. This is often evidenced by their drawings, such as the one shown in Figure 2, which indicates that the student is aware of the relative sizes of the two sets involved. However, the equation generated is not viewed as an expression of equivalence, but rather as a description of relative size. Clem-

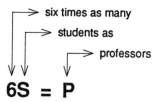

Figure 1. Syntactic translation of the student-professor problem into an equation.

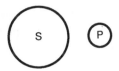

Figure 2. Student's drawing of the meaning of the student-professor problem.

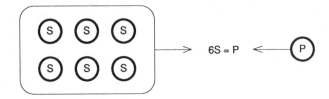

Figure 3. Semantic translation of the student-professor problem into an equation.

ent (1982) points out that the letter S is not perceived as a variable representing the number of students, nor is the equal sign used to express an equivalence, but rather a comparison or association (see Figure 3). Whether the cause of the error is syntactic or semantic, it is quite obvious that, in both cases, natural language interferes with the translation into algebra.

One might also argue that these errors are due to the fact that the problems involve proportional reasoning and that this is being ignored by researchers in the examination of their data. Kaput (1982) did attend to this possibility in his study of the student-professor problem and found that, in addition to using $6S$ and P in reversed equations, students also used these terms in algebraic statements such as $6S + P$, as well as in proportional reasoning statements such as $6S/1P$ and $6S{:}1P$. As with the reversed equations, the 6 is used as an adjective qualifying S, which is regarded as a noun—the concatenated 6 acting as a modifier rather than as a coefficient of multiplication. Kaput refers to such appearances of $6S$ as the "adjectival-nominalist" error and suggests that the same linguistic/cognitive mechanisms also underlie the reversal mistakes. In examining the cognitive obstacles involved, he notes that various levels of proportional reasoning competence interact with fundamental misconceptions (as expressed in the students' mistakes) and with a poor understanding of mathematical variables. Kaput also notes that some of the traditional prealgebraic representations of ratio and proportion, particularly the $A{:}B{::}C{:}D$ forms, do not easily translate to correct algebraic equations.

Attempted Remediation

Rosnick and Clement (1980) have explored the effects of tutoring strategies on these algebraic misconceptions. They worked individually with nine students, most of whom had taken one semester of calculus and all of whom had initially shown reversals in the student-professor type of problem. In taped interviews, they attempted various remediation strategies: simply telling the students that the reversal is incorrect; stressing that the variable should be thought of as the *number of students*, not just *students*; pointing out with pictures that, since the group of students is larger, one must multiply the number of professors by 6 to establish an equality; asking the students to test their equations by substituting numerical values; asking them to draw graphs and/or tables; showing them how to set up a proportion in order to solve the problem; and demonstrating a correct solution in an analogous problem. Based on an analysis of their interviews, the authors concluded:

> The reversal problem is a resilient one and . . . students' misconceptions pertaining to equation and variable are not quickly "taught" away; in fact, at least seven out of the nine students demonstrated in one way or another that they maintained the reversal misconception. (Rosnick & Clement, 1980, p. 6)

Wollman (1983) studied the effect of instructional strategies using numerical computation and comparison of quantities on the rate of correct translations from sentence to algebraic equation. He reported that after only 10 minutes and without the benefit of a tutor for prompting, 16 of the 17 students who initially erred were able to arrive at a correct equation. Following a critique of his paper (Kaput, Sims-Knight, & Clement, 1985), Wollman (1985) did agree with his critics that the students could have used a checking procedure to verify the correctness of an equation regardless of whether they understood how to generate the correct equation. Hence, their improved performance cannot be taken as an indication of improved understanding.

The danger with remediation based on monitoring one's answers is that it can lead to improved performance and that such positive results may reinforce the instructor's tendency to ignore the genuine cognitive obstacles at hand. Cooper (1984) reported some remediation that used a teaching unit on direct variation. The notion of proportionality was discussed, as well as the transformation of a proportionality statement $(A \propto B)$ into one of equality $(A = kB)$. Based on his students' improved performance, Cooper took issue with Rosnick and Clement's conclusions. Unlike Wollman, Cooper used instruction in his attempted remediation—instruction rightly emphasizing proportional reasoning. However, Cooper's disagreement with Rosnick and Clement was based on his students' test performance, and not on the quality of their thinking. Students' thinking processes can be adequately assessed only through individual interviews.

The translation problems that have been studied by so many researchers may not prove to be of capital importance in themselves. However, they have provided an opportunity to investigate the deeper roots of the problems, to uncover some of the cognitive obstacles involved. Kaput (1982) summed it up well when he pointed out that students having difficulties with these translations

> are unable to overcome the infiltration of natural language symbol-referent patterns and natural language syntax into their algebraic representations (p. 1); teaching and learning mathematics . . . involves . . . a rather complex interaction between highly stable old knowledge structures and permanent "natural" linguistic mechanisms on one hand, and new knowledge structures and symbol systems on the other. (pp. 8-9)

GRAPHS OF EQUATIONS IN TWO VARIABLES

A major leap in the mathematical development of high school students occurs when they are introduced to the notion of the graph of an equation in two variables. That the majority of students do not manage this leap is substantiated by NAEP results. Carpenter et al. (1981) found that, although students could graph ordered pairs in the Cartesian plane, most did not seem aware of the relationship between equations and their graphs, even for simple linear equations:

- When given a ruler and a sheet of paper with labeled axes, only 18% of the 17-year-olds tested were able to produce a correct graph corresponding to a linear equation. Moreover, about 25% of the students with one year of algebra and less than 50% of those with two years were successful.

- The reversal of the problem proved to be even more difficult. Given a graph of a straight line with indicated intercepts (-3,0) and (0,5), only 5% of the 17-year-olds and about 20% of the students with two years of algebra could write the equation.

These results indicate that the failure to understand graphs is widespread. Yet, many teachers find it hard to see why a topic that seems to be a logical extension of the work with equations proves to be so difficult for the learner. They believe that, by linking algebra to geometry, the added visual dimension should make it easier for students to understand algebra in two variables. The difficulties students encounter are caused by cognitive obstacles that we are just beginning to discover. We have evidence of their existence but are still unable to identify them specifically.

The Construction of Axes and Scales

It is easy to underestimate the level of sophistication involved in the construction of the Cartesian plane schema. Since most students manage to provide the coordinates of points in the plane and, conversely, locate points when given their coordinates, we conclude that they generally understand the structure involved, save for a frequently occurring reversal of the coordinates. Yet, there is evidence suggesting that even the notions of axis and scale are more difficult than expected. Vergnaud and Errecalde (1980) investigated the translation of numerical data onto a straight line by 150 students in the last grade of primary school and the first two grades of secondary school. Four different types of problems were used: weights of babies, ages of young children, dates of birth, and javelin throws. The protocols of the interviews indicated two main streams of thinking: students who ordered the problem data as dots on a scale and students who represented magnitudes as segments of a line. In the latter group a hierarchy was discovered:

1. At a fairly primitive level, some students used separate segments for different magnitudes and, hence, the notion of axis was not yet present.

2. The first presence of the notion of axis appeared among students who still used separate segments for different magnitudes but now put these end-to-end (see Figure 4).

3. The first recognition of the need to start from a common origin was shown in the work of students who drew different segments for different data but aligned their segments on the left (see Figure 5).

4. At this level students set different data on the same line but still found it necessary to explain that each datum was represented by a segment starting from the origin (see Figure 6).

Vergnaud and Errecalde reported that, before being able to use a scale, students had to develop a new way of thinking; that is, for a datum to be

Figure 4. End-to-end positioning of segments.

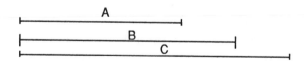

Figure 5. Left-end alignment of segments.

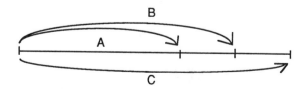

Figure 6. Left-end alignment of segments on the same line, with explicit representation of the starting and end points of each segment.

represented by a dot on a scale, students must map the segment onto its final endpoint. Before being able to translate data into dots on a scale, some students represented the data qualitatively as points on a line, the position of the points simply indicating the relative order of the quantities without representing their measure. Furthermore, the distances between the points on the line did not represent any measured difference between the given magnitudes. For instance, for javelin throws of $A = 68$, $B = 68.5$, and $C = 70$ metres, the data would be represented according to Figure 7.

For dots on a line to represent both the relative order and the actual measure, the added notion of an "interval scale" is essential. It is only when a line has been "graduated" by the iteration of a given interval that it can become an axis and, hence, a scale on which data can be represented both as points *and* segments.

This study reveals the sophistication involved in the construction of the concept of axis. While no comparable investigation has been carried out on the learning of the Cartesian plane, the mere fact that it is two-dimensional cannot but add to the complexity of constructing this concept.

Evidence of this can be obtained from a question asked by Kerslake (1981) in her investigation of students' understanding of graphs. She tested second-, third-, and fourth-year pupils in British secondary schools (13, 14, and 15 years of age). One of her questions dealt explicitly with the choice of axes and scales. Students were asked to plot the points $(20,15)$, $(-14,3)$, and $(5, -12)$ on a sheet of graph paper. The objective here was to find out whether students were able to select their own scale and position their axes so that negative numbers as well as positive numbers could be plotted. The results were quite startling: After two years of algebra, 38.7% of the 15-year-olds could not provide suitable scales and 34.6% could not correctly position their coordinate axes. In her interviews with some of the students, Kerslake discovered some cognitive obstacles. Some students believed that it was possible to have different scales for the positive and negative axes; others did not see the need to position their axes at the origin and used some other number; a third set of students simply ignored the negative numbers and did not provide for them on their scales. Although the second misconception can be viewed as a misunderstanding of the conventions to be followed, the other two misconceptions are clearly related to the construction of interval scales in two dimensions.

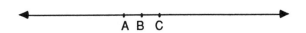

A B C

Figure 7. Position of points indicating relative order of quantities without representing their measure.

Transition to Continuous Graphs

In graphing equations in two variables, it is not enough for the learner to be able to relate points in the plane to ordered pairs of real numbers. A major cognitive step involves the notion of *continuity*—the graph of any polynomial function being continuous. But, for a student to grasp this idea, a continuous graph must be perceived as an infinite set of points. That most students do not perceive this is substantiated by the answers given to another question on the Kerslake test: Given a Cartesian plane, students were asked to plot some points and were told that these lay on a straight line that they were to draw. They were then asked, "How many points do you think lie on the line altogether?" Barely one fifth (19.6%) of the 15-year-old students perceived the straight line as an infinite set of points. About 10% of them answered that there were "hundreds" or "lots," while 51.6% thought that there were a finite number of points (e.g., 4, 5, 8, etc.).

Kerslake's interviews indicated that pupils found it difficult to conceive of points on the line other than those they had plotted. Some students gave the number of points that had integral coordinates and some counted the number of points where the line crossed the grid. These results seem to contradict those of Piaget and Inhelder (1947/1977) who found that, by the age of 11-12 years, Geneva children had abstracted the notions of *point* and *line*—perceiving a point as dimensionless and a line as an infinite set of points.

Some of the tasks used in the Piagetian study referred to above were included in a pretest to a teaching experiment involving three case studies carried out by Herscovics (1979a). The subjects were 15-year-old boys in their third year of secondary school who had not yet been exposed to equations in two variables, their graphs, or the Cartesian plane. One of the pretest questions dealt with the number of possible consecutive subdivisions of a given line segment—according to Piaget, the pupils were expected to perceive the number of subdivisions as unlimited and resulting in a point. Another pretest task asked how many points could be inserted between two given points, *A* and *B*, about 5 cm apart—the students were expected to realize that infinitely many points could be inserted and that the end product would be a line. Excerpts of protocols obtained from the pretest interview with Colin, the student considered to be strong in mathematics according to the school authorities, are quite revealing:

> Regarding the first task, Colin thinks the segment could be subdivided four or five times. Following Piaget's hint to consider the remaining piece as something that could be stretched (like a rubber band), the student agrees that the process can be repeated "continuously." However, he does not believe that the piece left at the end has any points but that "if you look under a microscope, . . . [it will have the shape] of a square" (Herscovics, 1979a, p. 95). . . . In response to a direct question, Colin states that he does not think of a line as a set of points. Regarding the task of inserting points between *A* and *B*, he thinks that he could insert about 30 of them, and if their size is reduced, about 68. But if they were even smaller he "could put thousands of them." He feels that

the result would "look . . . like a line," but in reality it would not be a "straight line," all "flat" as when drawn with a straightedge, because "a point is a mark" and this means "it has a center" and, hence, by inserting points one obtains a "*point-line*." He draws a set of points [see Figure 8] while stating that it "kind of waves" (indicating the upper part). (pp. 101-102)

Interviews with the other two students indicated that neither of them viewed straight lines as sets of points—let alone infinite sets—nor did they perceive points as dimensionless. Quite obviously, we cannot take for granted that such abstract levels of conceptualization have been achieved by novice learners. But neither must it be assumed that such levels are difficult to reach. For, in fact, the Piagetian tasks provided these students with an opportunity to think about these questions for the first time. And, in the subsequent interview, they all indicated that, even if they did not think that a line was an infinite set of points, at least they were quite sure that it could be "covered" by an infinite set of points. In the case of a finite line segment, one simply had to make the points "small enough." Thus, the notion of covering a line with points can be viewed as an intermediary step in crossing the point-line continuity hurdle.

Linear Equations and Their Graphs

For equations in two variables, the simplest possible graph is that of a linear equation. Yet, even at this level, the notion of graph presents major cognitive obstacles to most students. In an extensive assessment of British students, Kerslake (1981) found that, even when the question of graph was related to a real-world situation, it did not prove easier. Students were given the equation $W = 2H + 3$, where 3 was the basic wage in pounds and 2 was the hourly wage for overtime. They were asked to evaluate W for $H = 1$ and for $H = 2$, plot the information, and draw the graph of the equation on a grid marked with axes and scales. Among the 15-year-olds, 60% evaluated W, but only 35.8% plotted the points, and a mere 19.9% drew the line. Kerslake points out that it was surprising to find so many pupils unable to plot the points in this particular instance, since they had been able to plot points when given sets of ordered pairs. Thus, for novices, there seems to be a cognitive gap between being able to obtain pairs of numbers from an equation and transforming these numbers into pairs of coordinates.

Immediately after working on this problem, the students were asked to plot the graph of $y = 2x + 3$. Results indicated that it made little difference (a 3% drop) whether an equation originated from a relevant problem or not. The 16% success rate of these British students was similar to the rate obtained in the NAEP study mentioned earlier: 18% of the 17-year-old

Figure 8. Colin's drawing of a "point-line" showing how it "kind of waves."

American students tested were able to produce a correct graph of a linear equation. The similarity of the data can hardly be construed as mere coincidence.

Although the generation of the graph of a linear equation proves to be an arduous task for most students, they find it easier to recognize the graph of a linear equation, but harder to produce the equation of a given straight line, as evidenced by the results of the Kerslake test (1977, 1981). Her students were given the four graphs shown in Figure 9 and were asked: "Which of the graphs represents the line $y = 2x$? What are the equations of the other lines?"

That identifying the graph of $y = 2x$ was easier (26.9% of the 15-year-olds succeeded) than generating the equations of the other lines may be due to many reasons: It was the only equation that was given; the equations of horizontal and vertical lines prove to be more difficult since one of the variables is missing; equations of straight lines whose y-intercept is not zero also prove to be more difficult to produce than those of lines intersecting the origin (Herscovics, 1979b, 1980). As Kerslake (1977) notes, the graph problems given above are found in most British textbooks, very often in Years 1 and 2 of the secondary school. Thus, she found the results for the fourth graph of Figure 9 most disappointing: "Anyone brought up to pick out points on a given line and to look for a pattern connecting the coordinates should find [this graph] no more difficult than any other" (p. 24). Yet the success rate for that graph (5.1%) was again quite comparable to the success rate for a similar question in the NAEP study (5% of the American 17-year-olds could write the equation of a line with indicated intercepts). That two years of algebra raised the success rate to a mere 20% indicates the extent of the cognitive obstacles involved.

These obstacles vary for each student, and only through an individual interview are the specific hurdles likely to be uncovered. For many pupils, their problems can be traced back to a failure to construct an adequate interval-scale concept. For others, it may be the continuity gap in the transition from points to lines. For another group of students, it could be a weak

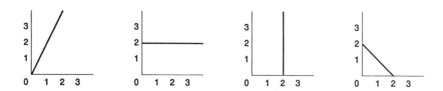

Figure 9. Task: "Which of the graphs represents the line $y = 2x$? What are the equations of the other lines?"

coordinate pair linkage between equations and graphs. For most students, their difficulties are probably due to a whole sequence of cognitive obstacles that they have never quite managed to overcome.

Interpretation of Graphs

The discussion of the cognitive obstacles associated with graphs would not be complete without reporting the research dealing with students' misinterpretation of graphs. A picture is supposed to be worth a thousand words, but this does not prove to be quite true for secondary students looking at graphs. Janvier (1978, 1981a) has found that they read graphs essentially in a "point-by-point" manner and ignore the more global features. He has observed that students are generally asked to plot a graph from a table of ordered pairs but are subsequently presented with questions that could more easily be tackled from the table alone. No wonder then that graphs are used more like tables than for the more general information they can display. Seldom before a course in calculus are students asked about extrema, intervals over which a function increases or decreases or levels off, discontinuities, rates of change, and so forth.

Kerslake (1977, 1981) points out that, in Britain, travel graphs are part of most courses on graphs. In general, students are given a distance-time graph representing a journey and are asked to identify different rates of travel, arrival time, and so on. In her test on the understanding of graphs, Kerslake presented students with graphs that could *not* represent journeys (see Figure 10) and asked them: "Which of the graphs represent journeys? Describe what happens in each case." She found that many pupils made no response at all. Of those who responded, many described the first graph as "going along, up, and along," "climbing a vertical wall," "going east, then north, then east," and so on. The third graph was often interpreted as "going uphill, then downhill, then up again" or "climbing a mountain." Her results indicated that few students were able to interpret these graphs appropriately and that many were led astray by purely visual configurations. This was also the case in the results reported by Janvier (1981b) who found that, in reading

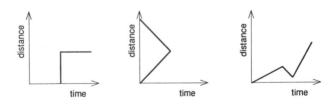

Figure 10. Task: "Which of the graphs represent journeys? Describe what happens in each case."

the graph representing the speed versus distance of a racing car along a track, students confused the graph with the track and, thus, were unable to read off the graph the number of corners on the track.

Other cognitive obstacles have been uncovered (Bell & Janvier, 1981; Janvier, 1981a; Ponte, 1985). One of these obstacles appears with questions on nonlinear graphs in which students are asked to identify an interval on which the largest increase occurs. Rather than focusing on the largest *increase*, students answer with the largest *value* of the function. Another difficulty with this type of question is students' tendency to provide a *point* response rather than an *interval*. Both of these responses indicate that a pointwise interpretation of graphs is deeply anchored in learners' cognition and prevents them from moving on to a more global perception.

FUNCTIONS

Few concepts in mathematics are as fundamental as the notion of function. This is the reason that functions have been greatly emphasized in the mathematics curriculum. In fact, even at the primary level, children are introduced to "function machines" with numerical inputs and outputs from which they have to discover the arithmetic operation performed by the machine. Whether or not the child is thereby dealing with the function concept is debatable, since the defining property of a *function rule*—the uniqueness of the output—is merely implicit. The same remark applies to most activities involving "function tables," equations in two variables, or their graphs. It is tempting to claim that, when students work with these mathematical objects, they are also working with the notion of function. However, unless the learner has some awareness of the underlying notion of function that is involved, the related activities can only be viewed as providing an intuitive basis for the eventual construction of the function schema.

NAEP Results

Among the tasks used in the NAEP assessment (Carpenter et al., 1981), some dealt with the evaluation of functions, others with the completion of a function table, and still others with the interpretation of a formula in two variables. Thus, the level of comprehension investigated was essentially that of an intuitive understanding of the function concept. The first question, one dealing with evaluation, was the exception, since its second part included formal function notation: Students were asked to evaluate $a + 7$ when $a = 5$, and then $f(5)$ when $f(a) = a + 7$. Although nearly all students (98%) could evaluate $a + 7$, it is impossible to know if they perceived $a + 7$ as a function. But it is clear that functional notation introduces new difficulties, even for students with two years of algebra, since the success rate for the second evaluation dropped from 98% to 65%. Functional notation is very efficient because it condenses a great deal of information, but this economy of notation can also create a cognitive obstacle.

The task of completing a function table is based on the recognition of some pattern. Given two variables and a few values for each, the student has to discover a relationship between the variables. In one NAEP task testing this ability, students were asked to fill in the table shown in Figure 11. Most of the students with one or two years of algebra could recognize the pattern—adding 7—from the given numerical values (success rates of 69% and 81%, respectively). However, about a quarter of those who succeeded with the numerical part of the table (28% and 23%, respectively) were unable to generate the equation $y = x + 7$. These results indicate that the obstacles involved in producing an algebraic equation in two variables are not restricted to translating word problems into equations. Even when the rule relating two variables has been discovered, many students cannot formalize it into an algebraic expression.

One could argue that the task of completing a function table involves the notion of function in an abstract way, that is, devoid of the kind of meaning associated with relationships originating from the environment and the natural sciences. Perhaps relating the variables to more relevant situations might reveal a better understanding. This possibility was investigated in a task dealing with a function expressed by a formula. In this problem, students were provided with the equation $W = 17 + 5A$, relating the weight of young boys to their age, and were asked, "According to this formula, for each year older a boy gets, how much more should he weigh?" They were given the choice of answering 5, 17, 22, or "I don't know." Quite surprisingly, formal instruction in algebra seems to have had little effect on success rates (54% for 13-year-olds vs. 58% and 64% for 17-year-olds with one or two years of algebra, respectively). Of those students with two years of algebra experience, 30% did not even bother to substitute values for A; they simply read off 17 in the formula or added the two numbers (17 and 5). Since these students apparently did not perceive the relationship between the variables, one has to question seriously whether they viewed the formula as a function, even in the most general sense.

Other Studies Dealing with Intuitive Notions of Function

The NAEP results show that, even at the intuitive level, more than one

X	1	3	4	7	n
Y	8		11	14	

Figure 11. NAEP task on completing the function table and generating an equation.

third of high school students with two years of algebra cannot handle simple problems on functions. But, even for college students, the use and role of literal symbols as functional variables is difficult to master in the translation from table to equation. Clement, Lochhead, and Monk (1981) gave a sample of 34 engineering students a weight-stretch problem in which a numerical relationship between the variables was provided in table form. The numbers indicated a 3 cm stretch for every 100 gm of weight, from which the students were to write an equation. Only 16 students (47%) gave a correct answer. Of the remaining 18 students, 7 made reversal errors, 6 made various other errors, and 5 gave no answer at all. These results confirm those obtained by NAEP: Even when a numerical relationship expressed in table form is quite simple, many students will not be able to express it correctly using algebraic notation.

Little research has been specifically oriented towards identifying the cognitive obstacles involved at the intuitive level of the function concept. Wagner (1981) found that algebra students are often unaware of the arbitrariness of the letters chosen to represent functional variables. Consequently, they believe that a change of literal symbol may induce a change in the values of a function table. She showed students a table headed with the variables B and C and asked them about the effect of changing C to A. Of the 14 students who had been exposed to some formal algebra, 4 did not "conserve function"—they believed that a change of letters could result in a change of value of the function. For most secondary school teachers, it must be quite surprising that something as simple as the alphabetical ordering of letters can create a cognitive obstacle that interferes with students' understanding of the arbitrary nature of literal symbols.

Another study dealing with the representation of intuitive functional concepts was carried out by Dreyfus and Eisenberg (1982) with classes of students from the sixth to the ninth grade. None of these classes had studied the concept of function at the time they were tested. However, the pupils in Grades 8 and 9 had covered a unit on Cartesian coordinate systems in Grade 7. The function-related concepts investigated were those of image, preimage, extrema, growth, and slope. Each concept was presented in three different settings—in the context of a diagram, a graph, and a table. Representing these concepts in the form of diagrams is quite difficult; hence, it is not surprising that the diagram representation resulted in slightly poorer performance than the tabular and graphical representations. Little difference in success rates could be noticed between the latter two. Nevertheless, students of higher ability and socio-economic background preferred the graphical representation of the concepts covered, while pupils of lower ability and socio-economic background preferred the tabular representation for image, preimage, growth, and extrema, but preferred the graphical representation for slope. These results seem to indicate that a majority of students may not be ready to work at the graphical level and that extensive work on functions in tabular form ought to precede work on graphs. Of course, even then, the cognitive obstacles involved in the translation from table to graph need to be carefully investigated.

Learning the Formal Concept of Function

The previous section dealt with some intuitive aspects of the function concept and described students' knowledge in the absence of any instructional considerations. The impact of a formal approach to the teaching of functions was investigated in the late sixties at the height of the "new math" movement. Although this research may seem somewhat dated, it is nevertheless relevant since, even today, most teachers still expect students to grasp the set-theoretic definition of function that continues to appear in most textbooks (Herscovics, 1982).

In a doctoral dissertation on stages in the attainment of the function concept, Thomas (1969) tested 201 seventh and eighth graders of above-average ability (IQ: 125 or above). Starting in Grade 7, the notion of function had been introduced as a mapping between sets using arrow diagrams, rules, ordered pairs, and graphs. A *function* was defined as an ordered triplet (F,A,B), such that F is a subset of the set of ordered pairs with first elements in A and second elements in B, and such that each element of A appears as the first element of exactly one ordered pair in F (Thomas, 1971a, 1971b). Test results suggested a hierarchy of four levels of understanding of the function concept. Considering that careful instruction aimed at these very able students resulted in 164 of them (82%) barely achieving the lowest level of comprehension, Thomas concluded:

> It was . . . a shock to this investigator to find that, in a group of students who had supposedly been carefully introduced to the concept of function, many could not distinguish functions from non-functions in simple and concrete situations. At the same time these students could carry out many of the processes associated with the function concept. One might speculate on this basis as to whether students should be allowed to work with the processes associated with functions and only later learn to discriminate sharply those objects that are functions. This has, indeed, been a traditional route. Current thinking, however, runs counter to this approach. (Thomas, 1971b, p. 26)

A similar study was carried out by Orton in Great Britain (Orton, 1970) with 72 subjects, 12 to 17 years of age, from the upper half of the mathematical ability range. It is quite remarkable that, after four years of instruction, only 52% of these gifted students achieved a level of understanding considered highest in Orton's hierarchy. Similar results were reported by Harrison (1973) who described a replication of the Orton study carried out by Anderson (1971) with 72 upper-ability high school students in Calgary. Harrison concluded that:

> The patterns in the findings do not seem to suggest that the three years of formal work with functions prior to grade 11 might be time well spent. Perhaps a better use of time

would be to have students explore function ideas in very concrete contexts . . . building a firm intuitive foundation for the later formal definition. (Harrison, 1973, p. 114)

If one recalls that the students in these three studies were judged to be above average, it is safe to predict that, with a more general population of students, results would be drastically worse. But these investigations are important, for they demonstrate that we cannot bypass learners' gradual construction of the function concept by initially presenting them with the end product, that is, with a set-theoretic definition of function. This is what Karplus (1979) had in mind when he remarked that virtually all the examples used in the above studies lacked a context that might have provided an intuitive basis for a functional relationship.

Students' Interpretations of Function

In Israel, the concept of function is introduced in ninth-grade mathematics classes through a formal set-theoretic definition—a many-to-one correspondence between elements of a domain and range (Markovits, Eylon, & Bruckheimer, 1983). However, functions are also introduced in science classes as a relationship between variables. Markovits et al. studied the effect of context (mathematics vs. science) on problems in which students were asked to draw the graph of a function by connecting given noncollinear points in a Cartesian plane. Two groups of ninth-grade students—a high-ability group and a low-ability group—were tested. Results indicated that most students provided a *linear* response, that is, almost all the graphs were composed of straight line segments. Interestingly, the context had an effect on the success rate. High-ability students were more successful with the pure mathematics problems than they were with the problems embedded in a scientific context. This trend was reversed for the lower-ability group. The main conclusion reported by the authors of this research was that many ninth graders "hold a linear prototypical image of functions" (p. 276). Markovits et al. believe that this is due to the fact that students view linear functions as being the simplest functions. The results of this study regarding students' "linear function" images are in accord with similar observations made by Karplus (1979). These findings illustrate how, inevitably, knowledge of linear functions becomes an obstacle in the later construction of a more general function schema.

Another important investigation has revealed the perception of function in the upper grades of secondary school. Vinner (1983) tested students in Grades 10 and 11 enrolled in two academically selective high schools. Although they had met the notion of function in ninth grade, they were formally introduced to it in the tenth grade as if they had never seen it before. The definition used was a set-theoretic one. Students' answers to the test questions indicated that, despite the formal definition, students had definite opinions as to what constitutes a function and what does not. Many students believed that:

- A function should be given by one rule. If two rules are given for two disjoint domains, then two functions are involved.
- A function can be given by several rules relating to disjoint domains, provided these domains are half-lines or intervals. But a correspondence with an isolated point in its range is not considered a function.
- Functions that are not algebraic exist only if mathematicians officially recognize them (by giving them a name or denoting them by specific symbols).
- A graph of a function should be "reasonable," that is, it should be "regular"—it should not have angles.
- A function is a one-to-one correspondence.

These students' perceptions of what constitutes a function could be dismissed as just so many misconceptions, that is, errors in their conceptualizations. However, the prevalence of these misconceptions militates against their dismissal as simple mistakes. On the contrary, these errors are all strong indicators of the existence of cognitive obstacles. Indeed, all of these misperceptions are due to an overgeneralization of the initial examples used in the introduction of the function concept. These overgeneralizations, which occur naturally, become hurdles in the construction of the more global notion of function.

The Vinner study of high school students' conceptions of function was expanded to include college students and junior high school teachers (Dreyfus & Vinner, 1982; Vinner & Dreyfus, in press). Five groups of students were tested: (a) college students taking a low-level course in mathematics, (b) students taking intermediate-level courses, (c) students in high-level courses, (d) mathematics majors, and (e) junior high school teachers. A supplementary question was added to the previous questionnaire: "What is a function, in your opinion?" Responses to this last item were quite revealing. The results indicated that, despite the stress on a set-theoretic definition of the function concept, the great majority of students did *not* accept it. As was hinted by the Markovits et al. study, the students were greatly influenced by their experience with functions in their science courses. The data showed that the first three groups of students opted overwhelmingly (73%, 70%, and 63%, respectively) for a scientific interpretation (a function is a dependence relation, a rule, an operation, a formula, and so on). Even the mathematics majors were about equally split between a scientific and a formal, mathematical interpretation. Only the teachers showed a strong preference (73%) for a formal interpretation. These results seem to indicate that a set-theoretic approach is not very successful and could even be considered counterintuitive. The value of a set-theoretic approach has also been questioned by several mathematicians and mathematics educators.

In a short historical review, Malik (1980) traces the definition of function as it has evolved from the time of Euler (an expression or formula representing a relation between variables), through Dirichlet (y is a function of x if there is a rule determining y uniquely for a given value of x), Caratheodory (a rule of correspondence between real sets), and the Bourbaki school (a subset of the Cartesian product of two sets). Malik points out that Euler's definition covers all the functions needed for a course in calculus or precalculus, that the Dirichlet wording is well-suited for a first course in analysis, and that the set-theoretic definition could be left until the study of topology.

Freudenthal (1973) has also taken issue with a set-theoretic definition. He points out, "Definitions should be operational; fundamental definitions that can be immediately forgotten are simply wrong" (p. 389). In comparing the classical and modern definitions, Freudenthal asserts:

> The set theory definition of function is by no means more exact than the original one [and] the reason why people prefer the definition of function via relations is a self-deception: they falsely believe that this approach better satisfies the extensionalist demands of modern mathematics. (p. 389)

Finally, the words of Boas (1981) are pertinent to the subject. In his parting message as editor of the *American Mathematical Monthly*, Boas raised the question, "Can we make mathematics intelligible?" and pointed out the need to take the audience into account:

> Authors of textbooks (lecturers too) need to remember that they are supposed to be addressing the students, not the teachers. What is a function? The textbook wants you to say something like "a rule which associates to each real number a uniquely specified real number" which certainly defines a function—but hardly in a way that students will comprehend. The point that "a definition is satisfactory only if the students understand it" was already made by Poincaré in 1909, but teachers of mathematics seem not to have paid much attention to it. (p. 727)

The reason for discussing at such great length the pros and cons of the various definitions of function is that it leads us toward the identification of two other sources of cognitive obstacles. That the set-theoretic definition constitutes a cognitive obstacle for many students is not too surprising since even the mathematical community does not readily accept this definition. Malik's paper shows that each new definition was the subject of heated debates among mathematicians and took many years to become accepted. Thus, here is an example of a cognitive obstacle that can be related to corresponding epistemological obstacles. In this case, learners' difficulties with functions are reflections of the history of the evolution of the concept.

Another source of cognitive obstacles can be elicited from the example of functions. As noted earlier, a set-theoretic approach failed to reach most students since, as observed by Karplus, such an introduction overlooks the examples needed to provide an intuitive basis for functional relationships. Thus, a set-theoretic presentation can be described as a *formal* one, in that it tries to create meaning for a mathematical object (a function) on the basis of a formal definition. The research reported here shows that most students experience difficulties with such a definition and, hence, do not accept it. Thus, one can also view the obstacles involved here as being *induced* by the formal approach, that is, by the type of instruction provided.

CONCLUSIONS

In the most general sense, cognitive obstacles can be identified with learning difficulties that are not of an idiosyncratic nature, but whose occurrence is widespread. Three distinct sources have been described in this survey: obstacles induced by instruction, obstacles of an epistemological nature, and obstacles associated with the learner's process of accommodation. Of course, this classification is not meant to convey a trichotomy, for many cognitive hurdles, such as those dealing with the function concept, can easily be ascribed to more than one source. However, only those that can be directly attributed to instruction are avoidable, while the others are inherent to the nature of the new knowledge to be acquired and to the learner's natural thinking processes.

The cognitive obstacles induced by instruction are often due to a formalistic presentation of the subject matter. In such cases, learners cannot relate the notion being introduced to their existing knowledge. The gap between the new material and their cognition is too wide to overcome. This cognitive discontinuity is a major obstacle that remains with them until they are provided with additional means to bridge it. Unfortunately, the existence of this gap is not always evident, for students learn to repeat definitions and manipulate symbols even if they are meaningless to them. It is often by a chance remark that their lack of understanding surfaces.

The historical study of any discipline or specific concept always uncovers epistemological obstacles that had to be overcome for any growth. There is little doubt that many of these epistemological obstacles have their parallel as cognitive obstacles encountered by the individual learner. However, this correspondence cannot be taken too literally since today's learning environments are significantly different from those of the past. For instance, the notion of positional notation is perceived by the average urban child before he or she can read or write, simply from looking around and by watching television. Although one can identify different levels of understanding in the kindergartner's construction of positional notation (Bergeron & Herscovics, 1986), the cognitive obstacles are quite different from the epistemological obstacles that were surmounted in the invention of this symbolization. Today's children need not reinvent this concept; they can simply reconstruct it.

The third class of obstacles, those associated with the learner's process of accommodation, are pedagogically the most challenging. No matter how much goodwill and care the teacher provides, these structural changes cannot be conveyed by mere transmission of information. Each learner is condemned to alter the mental structure in his or her own mind. What kind of pedagogical intervention can help the process along? Probably, the first step ought to be the introduction of some motivating factor. Problems need to be presented that can be understood by the learners but which cannot be solved within their existing knowledge (or at least not readily solved). Having created the need for change, the new material has to be organized into a constructivist teaching sequence, that is, a sequence starting from the learner's cognition and expanding from it. Of course, such a teaching sequence needs to take into account the cognitive obstacles that have been previously identified. For any mathematical concept that is new to learners, the best we can do is create conditions likely to enable them to complete the difficult process of accommodation. There are, however, no guaranteed recipes.

This survey has shown that an important body of past research has succeeded in discovering the existence of cognitive obstacles in the learning of algebra. Their existence has been deduced from the several assessment studies carried out in many countries. Their specific identification has been the result of individual interviews with students, which uncovered the hurdles they were confronting in a given learning sequence. Although this past research is significant, it barely scratches the surface. Quite clearly, much new research is still needed. But it need not limit itself to the identification of cognitive obstacles. One can also prepare teaching outlines, that is, sets of lessons aimed at overcoming specific obstacles. These teaching outlines can then be tried out in individualized teaching sessions in which the constant dialogue with the learner provides feedback. The investigator can thus assess if the teaching experiment has been successful in enabling the pupil to construct the new concept, and whether or not new obstacles have been introduced by the teaching outline. Such teaching experiments will eventually provide teachers with alternative presentations that teach to cognitive obstacles instead of ignoring them.

REFERENCES

Anderson, S. B. (1971). *The development of the concept of function in secondary school students*. Unpublished master's thesis, University of Calgary, Alberta, Canada.

Bachelard, G. (1983). *La formation de l'esprit scientifique* (12th ed.). Paris: Librairie philosophique J. Vrin. (Original work published 1938)

Bell, A., & Janvier, C. (1981). The interpretation of graphs representing situations. *For the Learning of Mathematics*, 2(1), 34- 42.

Bergeron, J. C., & Herscovics, N. (1986). The kindergartner's symbolization of numbers. In G. Lappan & R. Even (Eds.), *Proceedings of the Eighth Annual Meeting of PME-NA* (pp. 34-41). East Lansing: Michigan State University.

Boas, R. P. (1981). Can we make mathematics intelligible? *American Mathematical Monthly*, 88, 727-731.

Boyer, C. B. (1985). *A history of mathematics*. Princeton, NJ: Princeton University Press. (Original work published 1968)

Carpenter, T. P., Corbitt, M. K., Kepner, H. S., Jr., Lindquist, M. M., & Reys, R. E. (1981).

Results from the second mathematics assessment of the National Assessment of Educational Progress. Reston, VA: National Council of Teachers of Mathematics.

Chalouh, L., & Herscovics, N. (1988). Teaching algebraic expressions in a meaningful way. In A. F. Coxford (Ed.), *The ideas of algebra, K-12* (1988 Yearbook, pp. 33-42). Reston, VA: National Council of Teachers of Mathematics.

Clement, J. (1982). Algebra word problem solutions: Thought processes underlying a common misconception. *Journal for Research in Mathematics Education, 13*, 16-30.

Clement, J., Lochhead, J., & Monk, G. (1981). Translation difficulties in learning mathematics. *American Mathematical Monthly, 88*, 286-290.

Clement, J., Lochhead, J., & Soloway, E. (1979). *Translating between symbol systems: Isolating a common difficulty in solving algebra word problems*. Unpublished manuscript, University of Massachusetts at Amherst, Department of Physics and Astronomy, Cognitive Development Project.

Collis, K. F. (1974). *Cognitive development and mathematics learning*. Paper presented at the Psychology of Mathematics Workshop, Centre for Science Education, Chelsea College, London.

Cooper, M. (1984). The mathematical "reversal error" and attempts to correct it. In B. Southwell et al. (Eds.), *Proceedings of the Eighth International Conference for the Psychology of Mathematics Education* (pp. 163-171). Sydney, Australia: Mathematical Association of New South Wales.

Davis, R. B. (1975). Cognitive processes involved in solving simple algebraic equations. *Journal of Children's Mathematical Behavior, 1*(3), 7-35.

Dreyfus, T., & Eisenberg, T. (1982). Intuitive functional concepts: A baseline study on intuitions. *Journal for Research in Mathematics Education, 13*, 360-380.

Dreyfus, T., & Vinner, S. (1982). Some aspects of the function concept in college students and junior high school teachers. In A. Vermandel (Ed.), *Proceedings of the Sixth International Conference for the Psychology of Mathematics Education* (pp. 12-17). Antwerp, Belgium: Universitaire Instelling Antwerpen.

Freudenthal, H. (1973). *Mathematics as an educational task*. Dordrecht, The Netherlands: D. Reidel.

Harrison, B. D. (1973). Secondary school mathematics from a Piagetian point of view. *Mathematics teaching: The state of the art* (MCATA Monograph No. 2). Edmonton, Alberta, Canada: Mathematics Council of the Alberta Teachers Association.

Hart, K. M. (1981). *Children's understanding of mathematics: 11-16*. London: John Murray.

Herscovics, N. (1979a). *Compréhension de la droite et de son équation au niveau secondaire*. Unpublished doctoral dissertation, Université de Montréal, Montréal.

Herscovics, N. (1979b). The understanding of some algebraic concepts at the secondary level. In D. Tall (Ed.), *Proceedings of the Third International Conference for the Psychology of Mathematics Education* (pp. 92-107). Coventry, United Kingdom: University of Warwick.

Herscovics, N. (1980). Constructing meaning for linear equations: A problem of representation. *Recherches en didactique des mathématiques, 1*, 351-385.

Herscovics, N. (1982). Problems related to the understanding of functions. In G. van Barneveld & H. Krabbendam (Eds.), *Proceedings of the Conference on Functions* (pp. 67-84). Enschede, The Netherlands: National Institute for Curriculum Development.

Janvier, C. (1978). The teaching of graphs: A language approach. *Proceedings of the Thirtieth Conference of the Commission Internationale pour l'Etude et l'Amélioration de l'Enseignement des Mathématiques*. Santiago, Spain.

Janvier, C. (1981a). Difficulties related to the concept of variable presented graphically. In C. Comiti & G. Vergnaud (Eds.), *Proceedings of the Fifth International Conference for the Psychology of Mathematics Education* (pp. 189-192). Grenoble, France: Laboratoire I.M.A.G.

Janvier, C. (1981b). Use of situations in mathematics education. *Educational Studies in Mathematics, 12*, 113-122.

Kaput, J. J. (1982). *Intuitive attempts at algebraic representation of quantitative relationships*.

Paper presented at the annual meeting of the American Educational Research Association, New York.

Kaput, J. J., Sims-Knight, J. E., & Clement, J. (1985). Behavioral objections: A response to Wollman. *Journal for Research in Mathematics Education, 16*, 56-63.

Karplus, R. (1979). Continuous functions: Students' viewpoints. *European Journal of Science Education, 1*, 379-415.

Kerslake, D. (1977). The understanding of graphs. *Mathematics in School, 6*(2), 22-25.

Kerslake, D. (1981). Graphs. In K. M. Hart (Ed.), *Children's understanding of mathematics: 11-16* (pp. 120-136). London: John Murray.

Küchemann, D. (1978). Children's understanding of numerical variables. *Mathematics in School, 4*(7), 23-26.

Küchemann, D. (1981). Algebra. In K. M. Hart (Ed.), *Children's understanding of mathematics: 11-16* (pp. 102-119). London: John Murray.

Kuhn, T. S. (1970). *The structure of scientific revolutions* (2nd ed.). Chicago, IL: University of Chicago Press.

Lochhead, J. (1980). Faculty interpretations of simple algebraic statements: The professor's side of the equation. *Journal of Mathematical Behavior, 3*(1), 28-37.

Malik, M. A. (1980). Historical and pedagogical aspects of the definition of function. *International Journal of Mathematical Education in Science and Technology, 11*, 489-492.

Markovits, Z., Eylon, B. S., & Bruckheimer, M. (1983). Functions—linearity unconstrained. In R. Hershkowitz (Ed.), *Proceedings of the Seventh International Conference for the Psychology of Mathematics Education* (pp. 271-277). Rehovot, Israel: Weizmann Institute of Science.

Matz, M. (1979). *Towards a process model for high school algebra errors* (Working Paper No. 181). Cambridge: Massachusetts Institute of Technology, Artificial Intelligence Laboratory.

Mestre, J. P., & Lochhead, J. (1983). The variable-reversal error among five cultural groups. In J. C. Bergeron & N. Herscovics (Eds.), *Proceedings of the Fifth Annual Meeting of PME-NA* (pp. 181-188). Montreal, Quebec, Canada: Université de Montréal.

Orton, A. (1970). *A cross-sectional study of the development of the mathematical concept of a function in secondary school children of average and above average ability.* Unpublished master's thesis, University of Leeds, United Kingdom.

Piaget, J., & Inhelder, B. (1977). *La représentation de l'espace chez l'enfant.* Paris: Presses Universitaires de France. (Original work published 1947)

Ponte, J. (1985). Geometrical and numerical strategies in students' functional reasoning. In L. Streefland (Ed.), *Proceedings of the Ninth International Conference for the Psychology of Mathematics Education* (pp. 413-418). Utrecht, The Netherlands: State University of Utrecht.

Rosnick, P., & Clement, J. (1980). Learning without understanding: The effect of tutoring strategies on algebra misconceptions. *Journal of Mathematical Behavior, 3*(1), 3-27.

Thomas, H. L. (1969). *An analysis of stages in the attainment of a concept of function.* Unpublished doctoral dissertation, Columbia University, New York.

Thomas, H. L. (1971a, February). *The concept of function.* Paper presented at the annual meeting of the American Educational Research Association, New York.

Thomas, H. L. (1971b, April). *The concept of function.* Paper presented at the annual meeting of the National Council of Teachers of Mathematics, Anaheim, CA.

Vergnaud, G., & Errecalde, P. (1980). Some steps in the understanding and the use of scales and axis by 10-13 year-old students. In R. Karplus (Ed.), *Proceedings of the Fourth International Conference for the Psychology of Mathematics Education* (pp. 285-291). Berkeley: University of California.

Vinner, S. (1983). Concept definition, concept image and the notion of function. *International Journal of Mathematical Education in Science and Technology, 14*, 293-305.

Vinner, S., & Dreyfus, T. (in press). Concept images, concept definitions and some common mathematical behaviors with respect to the notion of function. *Journal for Research in Mathematics Education.*

Wagner, S. (1981). Conservation of equation and function under transformations of variable. *Journal for Research in Mathematics Education, 12,* 107-118.

Wollman, W. (1983). Determining the sources of error in a translation from sentence to equation. *Journal for Research in Mathematics Education, 14,* 169-181.

Wollman, W. (1985). A reply to Kaput, Sims-Knight, and Clement. *Journal for Research in Mathematics Education, 16,* 63-66.

Different Cognitive Obstacles in a Technological Paradigm
or
A Reaction to: "Cognitive Obstacles Encountered in the Learning of Algebra"[1]

David Tall
Mathematics Education Research Centre
University of Warwick

We stand at a point in history that has all the makings of a change in paradigm, in the sense of Kuhn (1970). This change is being caused by a major innovation in technology—the computer. Interestingly, this change is also occurring at a time when several different schools of thought are putting forward consonant theories about the difficulties of learning mathematics which may fruitfully be pulled together. Here I refer to such ideas as the theory of cognitive obstacles initiated in science by Bachelard (1938/1983) and now taken on by many contemporary French mathematics educators, the theory of cognitive "frames" featured in Davis (1984), the theory of concept definition and concept image in Tall and Vinner (1981) and Kaput's (this volume) "complex web of mental representations."

At this point in the Research Agenda Project conference on algebra, we are reviewing the research of the past before turning to the possible effects of the new technology. The paper by Nicolas Herscovics is restricted to the pre-computer curriculum, but in this reaction I intend to discuss two aspects: the internal one, which looks at the nature of the review, and the external one, which looks forward to implications in the new paradigm. It is therefore significant that the author selected as his focus of attention the notion of "cognitive obstacle," which appears throughout the pre-computer papers considered here and will no doubt continue to be a major feature of research in the future.

The papers discussed are mainly concerned with gathering and analysing information on the cognitive obstacles found in the current algebra curriculum, concentrating on three main topics: equations in two variables, graphs of equations in two variables, and the notion of function. The reviewer takes a standard school curriculum view of algebra, leavened with the wisdom of his experience in mathematical education. His viewpoint is essentially constructivist (though the term is not actually mentioned in the article), and he chooses to interpret the notion of cognitive obstacle in terms of Piagetian theory, where the learner is confronted with new ideas that cannot be fitted into the learner's existing cognition, leading to an inability to cope adequately with the new information. Some, but not all, of the research discussed has an explicit Piagetian foundation; in particular, the data found by the researchers are sometimes interpreted in terms of quasi-Piagetian stages, representing increasing levels of complexity that not all of the pupils manage to reach. In other studies there is simply a questionnaire presented, exploring likely phenomena; responses are then classified when the data are known.

The major problem with most of the investigations considered is that they are linked to the curriculum *as it is at present*, and one must consider to what degree the conclusions may remain relevant in a new computer paradigm. This consideration is intimately linked to the nature of two essentially different types of cognitive obstacles.

THE NATURE OF A COGNITIVE OBSTACLE

The notion of a cognitive obstacle was first introduced in the realm of science by Bachelard (1938/1983) and highlighted in mathematical education by Brousseau (1986). They have characterized an obstacle as a piece of knowledge that has in general been satisfactory for a time for solving certain problems, and so becomes anchored in the student's mind, but subsequently that knowledge proves to be inadequate and difficult to adapt when the student is faced with new problems. The implication of Piagetian stage theory is that there are certain *fundamental* obstacles that occur for us all. If such universal obstacles exist, they would therefore also apply in a new paradigm.

Obstacles based on topic sequence. I postulate that the reason for the belief in fundamental obstacles arises from the fact that certain concepts have a degree of complexity that makes it necessary to become acquainted with them in a certain order. For example, fractions are, of necessity, more complicated than whole numbers, and experience with operations on whole numbers leads to the implicit generalization that "multiplication makes bigger," which leads to a cognitive obstacle when the individual meets the multiplication of fractions less than one.

However, some topics that are traditionally taught in a certain order may not have the a priori property that one concept is essentially more complex than the other. For instance, fractions are usually met in traditional syllabuses before negative numbers, but there is no reason why, given an appropriate context, the two topics should not be taught in the reverse order.

One may hypothesise that cognitive obstacles are a product of students' previous experience and their internal processing of these experiences. Granted this hypothesis, it would follow that an alternative sequencing for the curriculum (where practicable) might change the nature of understanding and the type of cognitive obstacle that may arise.

For example, empirical research shows that the problem, "Multiply $3c$ by 5," is at a lower conceptual level than the problem, "For what values of a is $a + 3 > 7$?" (Küchemann, 1981). However, this difference in levels may be an artifact of a traditional approach to algebra in which manipulative skills are often taught before seemingly deeper conceptual ideas. In one experiment, Thomas (1988) used the computer to induce a conceptual understanding of the notion of a variable to experimental groups and compared results with the traditional algebra learning of parallel control groups. Findings showed that with the experimental pupils the traditionally accepted levels of difficulty were reversed (see Table 1).

Thus, the computer is likely to challenge many fondly held beliefs concerning the comparative difficulties of algebraic concepts. It may also help mathematics educators to sequence the curriculum in a manner more consonant with cognitive development.

Obstacles based on simple cases. In addition to the obstacles that seem to arise from topic sequencing, I contend that the way in which we limit the child to simple cases for a substantial period of time, before passing on to more complex cases, is bound to set up cognitive obstacles. Thus, simple cases of functions, limited to those given by simple formulae, will lead to the impression documented by Vinner and Dreyfus (in press) that a function cannot have two rules of correspondence. Similarly, graphs are always assumed to be continuous because the examples encountered by the student always have this property.

According to Hart (1983),

> The brain is by nature's design, an amazingly subtle and sensitive *pattern-detecting* apparatus (p. 60) . . . designed by evolution to deal with *natural complexity*, not neat "logical simplicities." (p. 76)

Thus our curricula, designed to present ideas in their logically simplest form, may actually *cause* cognitive obstacles, as for example, in the intuitive function concepts, wherein the student is asked if the change of literal symbol may induce a change in the values of a function table (Wagner, 1981). Here one recalls Brousseau's notion of a "didactical contract," the implicit, unspoken agreement between teacher and pupil as to the nature

Table 1
Contrast of Performance Between Computer Group and Non-Computer Group

Question	Computer group (experimental) % correct	Non-computer group (control) % correct
Multiply $3c$ by 5.	14	41
For what values of a is $a + 3 > 7$?	31	12

of the tasks to be carried out in the classroom. Looking through the Herscovics review, one is often struck by the difficulty in understanding the nature of the didactical contract in some of the questions as posed. Clearly the students do not understand the nature of the game that is being played nor the meanings of the algebraic symbols.

Already curricula are being designed that incorporate more complex problem-solving tasks from the outset, allowing the student to perform in a human way by abstracting relevant information from a rich context. The nature of cognitive obstacles produced by this type of approach is a matter for research.

Obstacles in a technological paradigm. The computer will allow topics to be approached in a richer variety of ways, allowing new sequences of ideas to avoid known cognitive obstacles, though computers are very likely to introduce new obstacles of their own. Witness, for example, the work of Nachmias and Linn (in press) who showed that a significant proportion of students interpret computer representations literally. In a physics experiment the students inserted a probe into a cooling liquid and the temperature was represented as a function of time in graph form on a computer screen. Unfortunately, the large pixel size onscreen made the graph appear jagged rather than smooth, which about one-third of the students interpreted as a true representation of cooling. They thought the liquid remained at a constant temperature for a while, then suddenly dropped a little (depicted by the onscreen fall to a lower pixel level). The "authority of the computer" may therefore be an impediment to learning, especially in the early stages.

On the other hand, the predictability of computer software may also be a powerful learning tool to help the student form meaningful mental representations of concepts currently known to provoke difficulties. For example, programming in both Logo (Sutherland, 1987) and BASIC (Tall & Thomas, 1986)—two very different languages—provides students with the possibility of a meaningful conceptualization of a variable and helps them perform better on standard test items involving the meaning of variables. This suggests that the cognitive obstacles that currently arise with regard to algebraic notation may occur to a different (and, one may hope, lesser) extent in the new computer paradigm.

AVENUES FOR FUTURE RESEARCH

The new technology points to interesting avenues for future research. First of all, in a changing paradigm, we need to ask the overall question:

- What role will algebra play in the new technological paradigm?

For example, the manipulation of algebra to solve equations will be less important for that class of problems for which a numerical solution is appropriate and for which simple numerical algorithms on the computer will suffice. Demana and Leitzel (1988) report how the early stages of algebra can be replaced by using a calculator to solve many real problems, including some that are very difficult to solve algebraically. Likewise, the existence of symbol manipulators may obviate the need to spend excessive time learning techniques that can now be carried out by a computer (though anyone who has used muMath or the new symbolic calculator, the HP28C, knows that it is vital to understand the essential algebraic principles, even if the computer is used to perform the algorithms). We must substantially rethink the role that algebra will play in the mathematics of the future, and this is no easy task.

The second fundamental question related to cognitive obstacles is how different kinds of computer experiences affect students' conceptualizations:

- How does the computer environment change the nature of mathematical concepts, the development of students' conceptualizations, and the related cognitive obstacles?

In view of the growing realization of the complexity of mathematical concepts, which cannot be explained to the learner purely in terms of mathematical definitions and logical development, we must also ask:

- How can we encourage students to participate *actively* in the construction of appropriate meanings, some of which will be very different in the future paradigm?

This latter question brings me to a point that is not explicitly germane to this review and that was rarely discussed in papers presented at the meeting: the question of student computer programming. A most valuable way of building and testing algebraic concepts is through programming, even though some concepts are bound to be bestowed with meanings different from those found in pencil-and-paper algebra. An elementary understanding of how the computer works, through programming, may provide insight into how variables are handled and how graphs are drawn, so that some of the cognitive obstacles mentioned so far may be confronted and discussed openly, thus enabling the student to construct richer and more coherent conceptualizations. In fact, programming the computer to produce solutions has been shown to improve students' understanding of the relationships among variables for a specific class of problems (Clement, Lochhead, & Soloway, 1980). Thus, the potential of programming as a means of dissipating some of the cognitive obstacles encountered in the learning of algebra deserves to be further explored.

FOOTNOTE

1. This paper was written in reaction to an earlier draft of the paper by Herscovics (this volume).

REFERENCES

Bachelard, G. (1983). *La formation de l'esprit scientifique.* Paris: J. Vrin. (First published 1938)

Brousseau, G. (1986). Fondements et méthodes de la didactique des mathématiques. *Recherches en didactique des mathématiques, 7*(2), 33-115.

Clement, J., Lochhead, J., & Soloway, E. (1980). *Positive effects of computer programming on students' understanding of variables and equations.* Amherst: University of Massachusetts, Department of Physics and Astronomy, Cognitive Development Project.

Davis, R. B. (1984). *Learning mathematics: The cognitive science approach to mathematics education.* Norwood, NJ: Ablex.

Demana, F., & Leitzel, J. R. (1988). Establishing fundamental concepts through numerical problem solving. In A. F. Coxford (Ed.), *The ideas of algebra, K-12* (1988 Yearbook, pp. 61-68). Reston, VA: National Council of Teachers of Mathematics.

Hart, L. A. (1983). *Human brain and human learning.* New York: Longman.

Küchemann, D. E. (1981). Algebra. In K. M. Hart (Ed.), *Children's understanding of mathematics: 11-16* (pp. 102-119). London: John Murray.

Kuhn, T. S. (1970). *The structure of scientific revolutions* (2nd edition). Chicago, IL: University of Chicago Press.

Nachmias, R., & Linn, M. C. (in press). Evaluations of science laboratory data: The role of computer-presented information. *Journal of Research in Science Teaching.*

Sutherland, R. (1987). A study of the use and understanding of algebra-related concepts within a Logo environment. In J. C. Bergeron, N. Herscovics, & C. Kieran (Eds.), *Proceedings of the Eleventh International Conference for the Psychology of Mathematics Education* (Vol. I, pp. 241-247). Montreal, Quebec, Canada: Université de Montréal.

Tall, D. O., & Thomas, M. O. J. (1986). The value of the computer in learning algebra concepts. *Proceedings of the Tenth International Conference for the Psychology of Mathematics Education* (pp. 313-318). London: University of London, Institute of Education.

Tall, D. O., & Vinner, S. (1981). Concept image and concept definition in mathematics, with particular reference to limits and continuity. *Educational Studies in Mathematics, 12,* 151-169.

Thomas, M. O. J. (1988). *A conceptual approach to the early learning of algebra using a computer.* Unpublished doctoral dissertation, University of Warwick, Coventry, England.

Vinner, S., & Dreyfus, T. (in press). Concept images, concept definitions and some common mathematical behaviors with respect to the notion of function. *Journal for Research in Mathematics Education.*

Wagner, S. (1981). Conservation of equation and function under transformations of variable. *Journal for Research in Mathematics Education, 12,* 107-118.

Cognitive Studies of Algebra Problem Solving and Learning

Seth Chaiklin

Center for Children and Technology
Bank Street College

The NCTM Position Paper on Research (National Council of Teachers of Mathematics, 1986) notes that "mathematics educators must assume leadership roles in conceptualizing, communicating, and implementing research related to mathematics instruction" (p. 20). In service of that goal, this chapter provides a perspective for understanding and interpreting the existing cognitive research concerning algebra problem solving and learning. To develop this perspective, I shall first present a model of the related processes of algebra problem solving and understanding, and then use this model to discuss the cognitive research on algebra learning. The analysis presumes that our educational interest is to help students acquire an *understanding* of algebra. This interest motivates a close attention to the student's ability to solve word problems, an ability which is taken as prima facie evidence that students have acquired an acceptable level of understanding. My review will focus primarily on word problem solving because most of the literature addresses this question, but cognitive research has also investigated equation solving, and this work will be examined as well.

There are two main themes in the perspective to be developed here. First, to solve algebra word problems successfully, students must be able to interpret and understand the mathematical relations in these problems. Though this point seems obvious to any mathematics instructor, researchers have carried out cognitive studies to demonstrate and document these phenomena and to elucidate some hypotheses about the components of understanding. In effect, these studies have made intuitive understanding quite explicit and, in doing so, have described exactly what students must learn to be successful algebra problem solvers. The second theme of the paper is that effective algebra problem solving depends on knowledge. An important goal for the psychological researcher is to characterize the content and organization of knowledge that enables successful problem solving. Although there are some general strategies that may be pertinent here, it appears that knowledge about concrete situations is an important component of effective problem solving. After reviewing the relevant literature about algebra problem solving and learning, I shall consider in general how research is applied in educational settings and make some specific remarks about the application of the research on algebra.

COGNITIVE PSYCHOLOGY AND PROBLEM SOLVING

The cognitive studies of algebra learning must be understood as an extension of a general cognitive model of problem solving. Therefore, before examining research results, let us briefly consider the typical research interests and theoretical perspectives that motivate cognitive studies of problem solving. These interests are often different from educational interests, and it is important to be aware of the differences in order to interpret appropriately the educational implications of the theoretical and empirical accomplishments of cognitive research on algebra performance and learning.

It is also easier to interpret particular experimental studies if one understands the general model from which they are motivated. Although specific research reports might not discuss this general model of problem solving in detail, it can usually be assumed that this model is being followed more or less explicitly.

Cognitive Model of Problem Solving

The goal of problem-solving research reflects the goal of cognitive psychology in general: to understand the sequence of mental operations and their products in performing a cognitive task. The ultimate aim is to understand the fundamental processes of thinking and learning that operate across all or several tasks, even though it is recognized that cognition must occur with specific content, such as algebra. There is a hope that we can find basic cognitive processes (such as short-term memory capacity and speed of information processing), as well as processes that underlie learning (such as general abilities to notice and construct conceptual relations). In both cases, these kinds of processes are sometimes called *operating characteristics* (Mayer, Larkin, & Kadane, 1983).

The focus on operating characteristics has not been prominent in the cognitive research on problem solving, though it has not been ignored. For example, the concept of *memory load*—the amount of information that must be attended to at any one moment—has received some attention in problem-solving research, but these studies have tended to consider the process of solving transformation puzzles (e.g., Kotovsky, Hayes, & Simon, 1985). The primary concern of problem-solving research has been to describe knowledge and skills in psychologically meaningful units that correspond plausibly to the thoughts and actions of problem solvers.

Cognitive studies of algebra problem solving have followed the general trend of problem-solving research, focusing on the structure and content of knowledge needed to solve algebra problems. A major contribution of these studies has been the development of theoretical and empirical techniques for analyzing the step-by-step processes involved in solving algebra problems. These techniques have enabled cognitive psychologists to make fine-grained descriptions of psychologically plausible organizations of knowledge to solve algebra problems. These structural descriptions specify the knowledge needed for successful problem solving and account for incorrect performance as the absence of knowledge or the presence of incorrect beliefs. Beyond some speculations that follow from a general cognitive model, cognitive psychologists have not directly addressed the relation between operating characteristics and the ability to solve or to learn to solve algebra problems.

Before discussing the research literature on algebra problem solving, let me introduce three main themes in the general theory of problem solving: performance, understanding, and learning.

Problem-Solving Performance

Most cognitive research on problem solving has focused on performance. Problem solving is generally considered to be a goal-directed sequence of cognitive and physical actions. The basic framework used to analyze problem solving describes knowledge in terms of *states* with *operators* making changes in those states. These states are sometimes called the *problem representation*. These theoretical constructs let us characterize problem solving in terms of representations and operations on representations. Correspondingly, the *outcome* of learning can be described in terms of the acquisition of new representational and operational abilities. This kind of psychological analysis is particularly useful in circumscribed domains like algebra word problems, where problem solving can be understood as a set of transformations on a problem representation to achieve a desired answer.

Applying this framework to problem solving, we can summarize the generic problem-solving process as follows: A problem solver reads a problem and forms an initial representation that usually includes the information to be discovered (i.e., the *goal*), as well as information considered relevant to the goal, such as the givens of the problem and any other equations or formulas that are known beforehand. This is a problem state. The problem solver then proceeds to apply operators to modify the initial representation of the problem by substituting given values into formulas, transforming equations, and introducing new equations. These cognitive actions leave the solver in a new problem state. The problem solver continues to apply operators until a state is achieved that the problem solver accepts as the goal state (or gives up, or runs out of time).

Understanding

This general approach to describing problem solving can also be applied to the analysis of understanding. The cognitive analysis of understanding centers on the representation process. Understanding is embodied in a constructed pattern of relations among concepts described in a problem. The main point is that understanding can be characterized with different levels of specificity (Greeno, 1983). For example, one person might represent only the specific objects in a problem, while another person would represent these objects as an instance of a more general principle.

Although we do not have a comprehensive account of algebraic understanding, we can identify two important characteristics that most educators would agree constitute an acceptable level of understanding. The first characteristic is the ability to translate a verbal description into physical terms. For example, a problem solver could draw a labeled diagram that describes the mathematical relations in a problem. The second characteristic is the ability to interpret a physical representation in terms of equations and operations, and vice versa.

These characteristics of understanding may be inseparable from effective problem solving. To obtain acceptable answers to problems, students must translate verbal problem statements into problem representations that veridically represent the entities, and relations among entities, described in the problem. For example, in rate-time-distance problems, a problem solver has to represent the multiplicative relationships among rate, time, and distance, if successful solutions are to be reliably achieved. If students can characterize the relationships in physical terms, then we would attribute to them a degree of understanding of the problem that should be sufficient to generate an adequate solution. Hence, a central feature of competent problem-solving skill hinges on this conceptual understanding. I shall discuss these representational processes further in the next major section.

Learning

Cognitive psychology research on algebra problem solving has focused on practical questions of how students *perform* algebra tasks rather than on more difficult theoretical questions of how students *learn* to be effective algebra problem solvers. No doubt, having a description of the psychological components of algebra problem solving and identifying the points where students have difficulties is important; however, such a description does not necessarily provide a guide for understanding how students acquire these abilities. Cognitive studies of algebra learning tend to examine issues from the point of view of cognitive performance models (e.g., how students acquire problem-solving operators and representational abilities).

Significance of the Cognitive Approach

Before discussing the application of this general approach to the analysis of algebra problem solving, a word about the significance of this cognitive approach is needed. To some teachers, this structure-process form of analysis may seem stiff and abstract and unrelated to their experiences with the children they encounter in school. The use of abstractions and the attention to functional relations is not an attempt to escape from a difficult problem. Most of us have no trouble accepting abstractions in other scientific analyses. (My favorite is Feynman's analysis of "dry" water, in which viscosity is ignored in the analysis of water flow.) Similarly, we give functional descriptions of processes all the time, using abstractions from the physical details.

For example, we describe the operation of an elevator in terms of pushing buttons, the car moving to different floors, the door opening and closing. We do not doubt that these are "real" parts of the elevator's performance, nor do we discount this description because it does not include a discussion of the electrical and mechanical systems that underlie this functional description.

Along the same lines, there are often concerns about how to interpret the significance of the representations and operations proposed in a cognitive analysis. Do they really exist? Are they really in the head? Is this the form in which they are represented? For our purposes, the important theme is the *content* of the representations and cognitive operations—what referents are identified from reading the problem, what beliefs are held about these referents, and what operations are possible given a problem representation. At this stage in the development of cognitive theory, the physical existence of proposed cognitive structures is not particularly relevant. We do not need to claim that we are faithfully reproducing all aspects of human algebra performance. Just as we do not expect the planets to "know" the mathematical descriptions by which they behave, we do not have to postulate that our students explicitly "know" the mental operations that they follow. In both cases, we think it is important to have descriptions of these operations as an important step in understanding and explaining these actions.

I shall now proceed with a functional description of the process of solving algebra word problems, referring to published articles that support and illuminate this model. We can reflect on this working model, once it has been developed, and consider its implications for algebra learning and instruction. We should find that, even with the limitations of this model, the level of discussion about algebra performance and learning has been usefully elevated.

COGNITIVE MODEL OF ALGEBRA WORD PROBLEM SOLVING

In plain language, the basic view is that a person reads an algebra problem, notices what seems to be important, and modifies this initial reading to produce a solution. We will refer to the noticing process as *problem comprehension* and the modifying part as *equation solving* (Hayes & Simon, 1974; Kintsch & Greeno, 1985; Mayer, 1982a; Paige & Simon, 1966; Reed, 1987). These two processes can be further subdivided, according to the generic process of problem solving outlined above.

The comprehension process involves (a) reading the problem, (b) forming a mental representation that interprets the information in the problem into objects with associated properties, (c) organizing the relations among those objects, and (d) representing the relations as equations. The mental representations that are formed along the way reflect the state of the problem.

The equation-solving process involves transforming the equation(s) produced in the first stage to produce an answer. This process usually involves the use of both *strategic* knowledge, which is used to plan operations, and *procedural* knowledge, which enables a person to execute the specific transformations required to produce solutions of equations. This analytic decomposition of the psychological processes used in solving algebra word problems also corresponds to the instructions typically given in algebra textbooks. Paige and Simon (1966) describe one set from a 1929 algebra textbook, and these suggestions remain current.

Cognitive studies follow this general model with remarkable uniformity. Most such studies can be interpreted as providing a further elaboration on specific characteristics of problem comprehension and equation-solving processes.

Problem Comprehension

The problem comprehension process provides the problem solver with the initial representation of the problem from which problem solving proceeds. This process clearly draws on knowledge beyond what is explicitly given in the problem. Algebra problem solvers do not simply make a literal mental copy of all the given problem information. Evidence for this claim is found in Mayer's (1982b) study of memory for word problems. He found systematic differences in which propositions in the problems were remembered, as well as distortions of the given information. We shall consider the specifics of these results later. For now, the important point is that problem solvers interpret the given information in the process of comprehending a problem.

At a gross level of analysis, there are two main approaches to problem comprehension. One can be called a direct-translation approach; the other, a principle-driven approach. These labels are not entirely accurate, so let us consider the processes to which they refer.

Direct-translation problem solving. Direct translation refers to a process that is often characterized by a phrase-by-phrase translation of the problem into variables and equations. Sometimes students must use semantic knowledge in order to formulate those equations (e.g., knowing the value of quarters, dimes, and nickels), but syntactic rules are usually adequate for identifying variables and relations among variables. The viability of this process has been supported by a theoretical study in which a computer program was supplied with syntactic rules that translated English words and phrases directly into algebraic expressions. This program, supplemented with some special-case knowledge, could solve a range of traditional algebra word problems involving one or more equations in one or more unknowns (Bobrow, 1968).

It seems that many students use this direct-translation approach. Evidence is reported by Paige and Simon (1966), Hinsley, Hayes, and Simon

(1977), and Reed (1984). In general, the evidence consists of observing students solve a problem phrase-by-phrase, trying to translate each phrase of the problem into algebraic notation.

While this kind of problem solving may be effective for solving problems that are typically given in textbooks, the robustness of this approach is severely limited, which lessens its acceptability as an educational objective. In particular, students who use a direct-translation approach to solving problems are less likely to notice contradictions in problem statements. In a small study, Paige and Simon (1966) found that students who used a direct-translation approach did not detect any physical impossibilities in a set of four problems that contained contradictions. In contrast, students who constructed integrated, physical representations were able to identify some contradictions. This ability to construct integrated representations seems to require the use of general principles in solving problems and thus reflects what we would call greater understanding of the problem. Let us now consider the use of general principles in problem solving in more detail.

Principle-driven problem solving. In contrast to a direct-translation approach, in which attention is directed to interpreting each separate phrase of a problem, the *principle-driven* approach uses a mathematical principle to comprehend and organize the variables and constants of a problem. For example, when competent algebra problem solvers read a problem that begins, "A riverboat travels downstream . . .," they are usually able to recognize that the problem will ask the solver to compute either a distance, a rate, or a time (Hinsley et al., 1977). They also know other relevant facts (e.g., there is an additive relation between the speed of the boat and the speed of the current, with the sign depending on whether the boat is going with or against the current). Cognitive scientists have developed several different kinds of theoretical constructs to specify these interpretive operations. *Schema* is an important construct that is used to describe the principles that people use to solve algebra problems.

Because the schema is such a central idea in the cognitive analysis of algebra word problem solving, let me develop the idea in more detail with a simple example. The example also illustrates how the schema describes the parts of the problem that the solver considers significant. This kind of performance is often referred to as being able to "perceive the underlying structure of the problem." This "perceptual" ability may not be comparable to a single visual glance but, instead, is composed of several analytic steps, perhaps more comparable to the integration of inputs in the visual cortex than to its phenomenological result.

Consider this arithmetic word problem:

> Jeff has 5 marbles. He gives 4 marbles to Jan. How many marbles does Jeff have left?

There are several ways to interpret this problem. Here is my thinking-aloud description of how I solved this problem:

> I form an image of Jeff possessing five objects. It does not matter to me whether the objects are marbles, doughnuts, or auto parts. [To some students, it might matter.] I also form an image of Jan primarily as a receptacle for some objects from Jeff. Finally, I imagine the transfer of four of Jeff's objects to Jan, and I count how many objects Jeff has left.

How did I know how to characterize the problem in the way I did? Why did I consider the quantity of objects to be significant, but not their kind? How did I know what operation to do once I set up the problem states? As educators, we know intuitively that these kinds of cognitive operations have to occur for successful problem solving.

The notion of schema has been used loosely at times, perhaps giving it a reputation for obfuscating as much as illuminating. However, when we restrict our attention to algebra problem solving, we can identify a fairly definite and consistent use of the term. A problem schema has three main components: (a) representation of states in the problem, (b) representation of relations among those states, and (c) procedures that enable the generation or modification of those states. These three components and their use are discussed in Greeno (1983).

In the simple word problem cited above, we can describe my interpretation of the problem in terms of a part-whole schema (Greeno, 1983). The *whole* is the total set of marbles and the *parts* are the two subsets. The relation among these sets is that the union of the parts must equal the whole. The procedures that allow me to operate with this structure are the numerical operations of addition and subtraction. For example, if I know the value for the whole and one of the parts, then I know a procedure to compute the value of the other part by subtracting the known part from the whole.

The main idea in principle-driven analysis is that problems are interpreted into a structure of schematic relations that have associated procedures for operating on the structure. These schematic relations constitute a person's understanding of a problem. This is not to say that problem solvers who use schemata will always produce correct answers. There is no guarantee that problem solvers will form schematic structures that incorporate either relevant relations or all the relations needed for a solution.

A major approach to analyzing word problem solving is guided by the theoretical model of discourse processing. Kintsch and Greeno (1985) provide a detailed development of this theory by applying a well-established model of general text processing (Van Dijk & Kintsch, 1983) to an analysis of the schematic structures needed to solve arithmetic word problems (Riley, Greeno, & Heller, 1983). Their theoretical analysis shows that these two empirically derived models provide a plausible account of how students translate verbal problems into "situation models" that can be used for problem solving. The general processes are comparable enough to algebra so that adaptation is straightforward, and Reed (1987) has already started in this direction.

There is plenty of evidence that high school and college students who are knowledgeable about algebra have schematic principles that they use to recognize and solve algebra problems. Hinsley et al. (1977) showed that these students were capable of categorizing problems into different classes of characteristic types, sometimes with as little information as the initial noun phrase of the problem. Moreover, the information that students used to categorize problems was information they used in solving the problems (e.g., pertinent equations and diagrams, appropriate procedures). This finding is consistent with the idea that schemata help to focus attention. Finally, Hinsley et al. showed that students could also categorize and solve "nonsense" problems (i.e., problems in which some content words are replaced with nonsense words). This result provides clear evidence that students can form conceptual relations among parts of a problem even when the parts refer to meaningless objects like "grix" and "sten."

Additional evidence shows the importance of schemata for focusing attention on the conceptual relations in algebra word problems. Mayer et al. (1983) reported that when students recalled word problems after studying them for two minutes each, they tended to make more errors in recall of irrelevant information than in recall of details relevant to the problem. This result is consistent with data from studies in the psychology of memory that show that some structuring device is necessary to retain information.

Finally, although competent students may be able to use schematic relations in solving algebra problems, students more typically have great difficulty in extracting conceptual relations from problems. When students are asked to recall algebra word problems that they have just studied, they are much more adept at remembering values assigned to variables than they are at remembering the relations among the variables (Mayer, 1982b). Also, students sometimes remember relations in a problem statement as assignments of values (e.g., the problem statement, "A rectangle is 4 inches longer than it is wide," might be recalled as "A rectangle is 2 inches wide and 4 inches long"), but they never convert assignment statements into relations. These results are consistent with the idea that forming conceptual relations among the parts of a problem is difficult. Further evidence for the difficulty in assigning relations can be found in Reed (1987).

Which one is used? It is not the case that algebra problem solvers use only one approach in solving problems. It appears that there is an interaction between the knowledge that a solver has and the approaches used to solve a problem. A problem solver will use special heuristics, if they are available, but will fall back on more general problem-solving strategies when special heuristics are not available (Hinsley et al., 1977; Reed, 1987). This important result reminds us that the psychologist's functional descriptions of algebra problem solving refer to possible approaches that an individual might use to solve problems, but we still have to understand the conditions under which different approaches might be used.

Equation Solving

After a problem solver has formed a representation of the problem with sufficient specificity that the relationships in the problem are represented in equations, then it is possible to use algebraic techniques to transform these equations into solutions. Studies of word problem solving do not usually examine the process of solving the equation, so we shall have to turn to studies that have looked explicitly at symbol manipulation. The transformation of algebraic equations that do not have semantic referents may or may not be the same as the transformation of equations that have been generated from semantic referents.

I have encountered cases in which students were capable of performing algebraic transformations on meaningless symbols, but could not perform the same operations on symbols that had physical referents. Therefore, generalizing results on pure equation solving to word-problem equation solving should be made tentatively. For present purposes, it is sufficient to note some of the main issues.

Equation solving involves the use of strategic and procedural knowledge. Strategic knowledge involves setting goals for which procedures to execute. A common set of strategic operators, identified analytically by Bundy (1975), are to "attract" instances of an unknown to one side, "collect" instances of a term by computationally combining them, and "isolate" an unknown by removing coefficients and powers.

Good strategic knowledge is tantamount to selecting efficient procedures for solving an equation. Lewis (1981) examined differences between experts and novices in algebra equation solving. He could not identify many differences. Experts were able to combine more procedural operations into a single step, but they did not display unusually good strategic knowledge nor did they avoid an occasional procedural mistake. Although experts and novices may have comparable skill in algebraic transformations of textbook problems, a more important question is whether experts and novices differ in the way they use equations to help comprehend problems. That is, the ability to manipulate equations does not imply a comparable ability to provide meaningful interpretations of those equations, a skill which usually distinguishes experts from beginners.

Several extensive studies have identified and catalogued errors that beginners make in solving algebra equations (Carry, Lewis, & Bernard, 1980; Matz, 1983; Sleeman, 1984). In general, these studies find that errors are often systematic, reflecting the problem solver's beliefs about what is to be done. For example, some students compute $(A + B)^2$ to be $A^2 + B^2$. This is often not a careless mistake. Rather, it reflects incorrect procedural beliefs on the part of the students. (For another perspective on the research dealing with equation solving, see the paper by Kieran in this volume.)

IMPLICATIONS OF COGNITIVE MODEL FOR LEARNING ALGEBRA

A cognitive approach to the analysis of learning algebra with understanding follows from the performance model of algebra problem solving just discussed. The present section explicates issues about learning that emerge from the cognitive performance model and considers what we currently know about these issues by reviewing some initial attempts at instructional intervention. As a rule, the instructional experiments conducted to date have not been particularly successful in helping students to develop word-problem-solving abilities. However, they illuminate the performance model and highlight issues that need to be addressed in developing effective instruction.

In the performance model of algebra problem solving, *learning* is the acquisition of schemata and the ability to apply them to different situations. In other words, according to the cognitive model, successful problem solving with understanding depends on being able to form adequate conceptual relations and knowing how to apply legal transformations to those relations. In addition to acquiring appropriate schemata, students must also be able to apply those schemata across a range of problems. This ability is usually called *transfer* because the student has transferred schemata from the specific contexts in which the ideas were learned to other related contexts. Thus, if I learned the part-whole schema with marbles, then I should be able to apply this schema to other objects as well. Apart from learning how to apply conceptual relations across a wide range of situations, it is useful to be able to apply legal transformations efficiently. This latter issue is sometimes referred to as *automation* of problem-solving operators. There is some evidence that automation of these operators can interact with the formation of schemata. This interaction is the crux of the relation between conceptual and procedural knowledge (e.g., Hiebert, 1986).

Schema Acquisition

Recalling the definition of a schema, *schema acquisition* "requires knowledge of (a) problem states, (b) the operators that can be used when a given problem state has been attained, and (c) the consequences of using particular operators" (Sweller & Cooper, 1985, p. 69). In other words, we want to understand how students learn to represent appropriate problem states, their relations, and operations on those states and relations.

Traditional classroom methods of instruction have included describing a general method of problem solving and providing students with lots of problems (e.g., the odd-numbered problems, when the book supplies the answers for the even-numbered ones), so that students can develop their skill and presumably acquire the appropriate schemata. In practice, we know that these methods have not been effective.

Cognitive studies of algebra learning have focused primarily on the acquisition and transfer of problem-solving schemata. The design of these studies is influenced naturally by the general model of algebra performance. Typically, students are presented with some examples to solve or study, are then provided with a set of related transfer problems, and finally, are tested on some aspect of their performance (e.g., speed or accuracy) to see what improved problem-solving ability results from the study problems. These sessions usually last no longer than one hour. It should be obvious that these studies may not represent optimal instructional conditions—neither in terms of time nor of technique. Moreover, since the psychologists who conduct these experiments are not usually experienced with classroom instruction of algebra, they may not always be a good source of ideas for organizing effective activities to communicate relations within and across problems. It would be a mistake, however, to dismiss these studies because they do not have a superficial correspondence to usual instructional conditions. The cognitive studies of algebra learning can illuminate the process of learning conceptual relations and the particular difficulties that students encounter.

As will be seen, there are three general points that emerge from the existing instructional studies done from a cognitive point of view. First, learning schematic relations for solving word problems is important, but it is often difficult for students, and it is not always clear in a given situation why one instructional method is more effective than another. Second, student problem solving and transfer can be sensitive to subtle, seemingly minor differences in the structure of algebra word problems. This suggests that the selection of problems is very important for determining the ability of students to notice relations. Third, careful analysis of the knowledge structures that are needed to solve different problems helps to account for the differential responses to problems, and it is especially this third point that illustrates the value of cognitive theory.

Conditions for schema acquisition. The previously discussed studies of memory for problem relations suggest that students have the most trouble comprehending the relational aspects of problems. According to theory, relational understanding depends on the presence of appropriate schemata. Learning studies have thus explored different methods for teaching students to extract the appropriate schemata. The more that instruction can communicate the relevant conceptual relations to the students, the more effective the instruction will be. At this point, we do not have a principled basis for saying why one method might be more effective than another for communicating schematic relations, as the following examples show.

Reed and Evans (1987) conducted three experiments that searched for an effective way to help students make good estimates of mixtures of different concentrations and quantities of a solution. They found that providing a written list of principles for solving these problems was more effective for improving answers on test questions than (a) providing students with examples of mixtures and their outcomes, (b) providing examples, but asking students to estimate before giving the outcomes, or (c) providing graphical representations of possible mixtures. However, the most effective method was to have students work with a carefully developed, closely matched analog to the mixture problem, one involving the familiar domain of mixing two quantities of water of different temperatures.

In an elaborate experimental design, Cooper and Sweller (1987) had higher-ability (best 8th-grade class) and lower-ability (2nd-best 8th-grade class) students either solve equations or alternate between studying a worked equation solution and solving a comparable equation for a short (4-problem) or long (12-problem) acquisition period. The motivation for these groups, which were designed to test the conditions under which worked examples facilitate performance, will be explained below. The main point, for now, is the differential pattern of effects that were obtained.

To test what the students had learned, Cooper and Sweller asked them to solve two similar problems and two transfer problems that were extensions of the acquisition problems. The lower-ability, short-acquisition group that had studied some worked examples solved similar test problems much faster and with fewer errors than the counterpart group that had only solved conventional problems. However, there was generally no difference between these two groups in either time-to-solution or number of errors when they had to solve two more complicated equations that used the same component operations. But, when these lower-ability groups were given the longer acquisition period, then the worked-example group usually solved the two transfer problems faster and with fewer errors. None of the differences that were detected for the lower-ability groups were seen with the higher-ability groups.

This pattern of results highlights a general point that it is unlikely that we will discover a single teaching technique that will be universally effective for all students. In this particular case, we see that lower-ability students needed a longer acquisition period for a difference to appear in what was learned from two different instructional methods. On the other hand, the amount of acquisition time did not make any difference for higher-ability students. These findings also point out the need to develop specific theoretical explanations of the factors that can affect schema acquisition.

Another example of the subtle manipulations that can affect learning are found in comparing the results of Reed, Dempster, and Ettinger (1985) with those of Cooper and Sweller (1987). Reed et al. found that providing college students with a solution to a mixture, work, or distance problem was not sufficient for helping students solve equivalent problems with different quantities. Cooper and Sweller, in an extension of Reed et al., found that they could obtain transfer on the distance problems if students studied more than one problem. Finally, Reed et al. found that, by studying elaborated solutions, students could transfer to corresponding problems but could not solve similar problems that required only a slight modification of the solution procedure. In general, students used a syntactic approach in which they attempted to substitute numerical values in the equation. Analysis of their errors showed the now familiar result that students had considerable difficulty in specifying relations among variables.

This collection of studies on schema acquisition illustrates the general point that there are many factors that can potentially affect the ability of students to acquire schematic relations with associated problem-solving operators. As Cooper and Sweller (1987) noted, "The subtle nature of the limitations that cause these differences points to the importance of using suitable subjects, procedures and materials" (p. 354). The particular patterns depend on interactions among the difficulty of the problems selected, the ability of the students, and the length of the study periods offered. We must be careful not to let labels like *high ability*, *long acquisition*, and *transfer problem*—which refer to situations that are defined relatively— obscure the important point that we must analyze the concrete particulars of an experiment to understand the particular patterns of interaction.

Sensitivity to problem structure. Although the following studies were not specifically designed to show that students are sensitive to structure, they illustrate this point nicely. Reed (1987) wanted to examine the factors that affect a student's ability to notice analogical relations between problems. Presumably, a student who knows how to solve one of the problems of the analog should be able to solve others. Reed constructed a set of problems using a 2 × 2 classification scheme of problems that had a similar or different story context and an isomorphic or slightly modified solution procedure. In the first study, he used cover stories that involved mixing quantities (i.e., wet mixture, dry mixture, simple interest). When college students were asked to rate how useful the solution of one mixture problem would be to the solution of another, Reed found that they had difficulty noticing isomorphic relations among problems with different cover stories, and they did not notice differences in solution procedures when the cover story was the same. He constructed another set of problems with work cover stories, using the same 2 × 2 classification. This time he found that students were much more successful at detecting isomorphic relations across cover stories and at noticing differences in solution procedures for similar cover stories. This ability to notice isomorphic relations for one kind of word problem (work), but not for another kind (mixture), highlights the general principle that analogical reasoning skill is not unitary. Students can be sensitive to underlying problem relations for some kinds of problems and not for others, even when these differences seem superficial. An important implication of this result, if it proves to be more general, is that students may not need specific training in detecting analogies. Rather, students may need assistance in understanding specific problem content (Reed, 1987).

Comparable differences were found when work and mixture problems were used analogically to solve another problem. In counterbalanced studies, college algebra students were given 10 minutes to construct correct equations for four work (or mixture) problems. Then they were given two minutes to study elaborated solutions to two work (or mixture) problems, followed by 10 more minutes to write correct equations for the problems originally posed. Two of the test problems were isomorphic to the study examples, and the other two were similar but had slightly modified solution procedures. Virtually none of the 45 students wrote correct equations for any of the problems during the first ten minutes. After studying the worked examples, the students showed marked improvement, with greater percentages of them writing correct equations for the isomorphic problems than for the slightly modified problems, and a greater percentage writing correct equations for the work problems than for the mixture problems. Here we can see that seemingly slight differences in problems can have a large effect (given the particular experimental constraints) on students' ability to construct correct equations.

Analysis of knowledge structures and processes. As we have just seen, many algebra problems that seem to require comparable knowledge and skill for solution actually are differentially difficult for students. In other words, common sense is not a reliable guide for predicting whether students will be able to identify and use the structure of typical algebra word problems. Close attention to specific knowledge structures and processes required for solving word problems, as well as the structural characteristics of the problems, can be very helpful in accounting for performance differences in tasks (see Mayer, 1981, for one analytic scheme).

The value of structure-process analysis is illustrated in Reed (1987), who provides a detailed propositional analysis of the conceptual relations needed to solve work problems compared to mixture problems, showing that the work problems he used had a simpler schematic structure than the mixture problems. This analysis, in conjunction with the attention to different kinds of analogical relations between a base and target problem, enabled Reed to account for the differential performance between work and mixture problems and the differential effectiveness of using different kinds of analogical problems. Analysis of the components of analogical processes motivated a further experiment in which Reed asked students to identify how the concepts in one work or mixture problem corresponded to the concepts in another work or mixture problem. Students were first given an example to study. Reed found that more correct mappings were made for isomorphic problems than for problems with slightly modified solution procedures, and more for work problems than for mixture problems. These results help to verify the cognitive analysis and show the importance of close attention to the structure of the problems.

One last point that should be noted from Reed's report is that in one case, for two problems, students were equally competent at constructing matching relations between problems, but they produced more correct equations for one of these problems than for the other. This result suggests that analogical matching may not be a sufficient condition for constructing a correct equation, a process which requires that relations between concepts also be constructed.

Another example of structure-process analysis comes from an interesting set of studies conducted by Sweller and Cooper (1985), who have investigated a familiar approach to helping students acquire problem-solving schemata for equation solving by giving students worked examples to study. In no case did the study of worked examples make the student performance worse than the technique of learning by solving conventional examples. This finding is significant because the students spent less study time on worked examples.

Sweller and Cooper's analysis and their arguments fit readily into the general model of algebra problem solving already presented. The essence of their argument is that problem solving is a goal-directed activity, and that students need to learn specific problem-solving actions in relation to goals. If students are taught to solve a particular kind of problem by simply solving large numbers of that kind of problem, then students will spend a great deal of time engaged in activities that do not expose them to the relevant conceptual relations that we want them to see. Consequently, Sweller and Cooper propose that students should study worked examples of problems as a method for focusing their attention on the relevant relations and operations to be learned.

Their proposal seems consistent with something that Herbert Simon often said in lectures: The real learning in problem solving occurs when you go back after you have solved the problem and try to understand what happened. Sweller and Cooper (1985) provided this kind of opportunity by highlighting the solution steps, which helped students figure out what to focus on. The basic experimental paradigm was to give students some solved problems to study, while the control group solved a comparable set of problems, and to give both groups a posttest to compare performance. Like the Reed studies, those of Sweller and Cooper reveal that (a) the use of worked examples is effective for helping students learn to solve equations that have the same structure but different variable names, but (b) the use of worked examples does not facilitate transfer to equations with a different structure that are solved with the same algebraic manipulations.

Their subsequent studies (Cooper & Sweller, 1987), described earlier in this paper, tested the hypothesis that a longer acquisition period should provide more opportunity for acquisition of schemata and automation of problem-solving operators. This prediction was partially supported. The lower-ability, worked-example group was able to make a significant improvement in their solution times and error rates on the transfer problems compared to the conventional group, which showed no change in their solution times. The higher-ability groups were already able to transfer.

Thinking-aloud protocols helped to illuminate what might be happening. Students, engaged in the same instructional procedures of worked examples or conventional problems, were asked to think aloud. The conventional problem-solving groups made more statements that were indicative of a means-ends analysis (e.g., "I'm trying to get the *a* equal to . . ."), while the worked-example group made statements that indicated a focus on schematic relations (e.g., "This problem is a bit like the ones I just did"). For the test problems, all the students made schematic statements for the similar problems, and they reverted to means-ends analysis for the transfer problems.

In sum, we see again the value of cognitive theory for illuminating and interpreting observed performance differences from different instructional conditions. Whether these fine-grained differences in performance make any difference in practice is a question that educators should address. Remember that the instructional methods may be less than ideal, so these studies are documenting the results of inadequate instruction. However, by identifying the specific knowledge structures and subject matter characteristics that affect the acquisition of problem-solving skill, we may learn what to address in our next instructional attempts.

DISCUSSION

We have reviewed a cognitive model of algebra word problem solving, considered empirical evidence that embodies the model, elaborated specific characteristics of algebra problem solvers in relation to the model, and examined some attempts to use the model for the analysis of algebra learning. I will now address the general question of how psychologically-oriented research can be applied to educational practices, and then close with some specific suggestions regarding the application of current research to algebra education.

Relation Between Research and Practice

The relation between research and practice is complex. Theoretically-oriented research is not usually translated directly into practice, at least not in education (e.g., Husén, 1984). In metaphoric terms, there is no pipeline between basic research and practice. Historically, psychological research has provided a rich perspective for interpreting the processes of learning and problem solving, which in turn provides guidance for curricular development and pedagogical practice (e.g., Committee on Fundamental Research Relevant to Education, 1977). Consequently, researchers and educators must consider the relationships between research and practice more explicitly. The implication for the research reviewed here is that it should not be viewed as immediately applicable to educational practice.

In light of this historical role of psychological research applied to education, I have tried to articulate explicitly the perspective that psychologists use to study algebra problem solving and learning. It is important to understand that research often proceeds in terms of what can be done, as opposed to what is needed. Therefore, some sensitivity will be required on the part of mathematics educators who are interested in using these research results to guide their practice.

In the next section, I would like to suggest some points where mathematics educators may have to take up interpretive work to translate the cognitive research on algebra into their practice of designing instructional activities. In accordance with the purpose of this chapter, I shall limit my remarks to suggestions within this cognitive perspective. In practice, educators may want to include other considerations in their designs, such as student interests and capabilities, as well as classroom constraints.

Implications for Educational Practice

I would like to distinguish between the *cognitive* difficulty of algebra problem solving and the *pedagogical* question of what we should do to help students with their cognitive difficulties. While the study of the cognitive aspects of problem solving has illuminated the kinds of intellectual difficulties that students face in solving these problems, it has not seriously addressed the pedagogical question of designing effective teaching practices and curricula. We still face the important challenge of how to organize material to capture and sustain interest so that students can engage in the intellectual processes identified by cognitive analysis as necessary or sufficient for acquiring knowledge of algebra. Similarly, we must develop units of study that correspond to manageable units for both the teacher and the student.

Algebraic skill. If there were any doubt, this review shows that algebra is not a singular skill or faculty—it is composed of myriad skills, many of them knowledge-based. Judging from the literature reviewed here, there is no reason to believe that algebra is an innate talent. Hence, when we refer to better or worse problem solvers, we refer to the level of performance in that task and not to an intrinsic characteristic of the problem solver.

Definitions of algebra. In general, cognitive research has not examined the question of what is to be counted as "algebra." Most of the studies have taken textbook problems as the definition of algebra. This is a nice illustration of a difference in interests between the psychologist and the educator. In the absence of any clear specification or widely accepted definition of the goals of algebra instruction, cognitive psychologists have simply selected the subject matter that they experienced in their algebra courses or found in current textbooks. I suspect that if educators communicated another view of the goals of algebra instruction, then psychologists would be willing to organize their studies around these issues.

Suggestions for Further Research

In this review I have argued that cognitive research has amplified, rather than redirected, current educational wisdom. Moreover, I suggest that cognitive research should continue in this same direction of amplifying what has preceded. Detailed cognitive analysis of the form and use of problem-solving schemata has been able to specify the psychological components that constitute knowing a schema. Also, the few attempts so far to help students acquire schemata have alerted all of us to the difficulties involved in helping students acquire these structures.

Mathematics educators should consider identifying the conceptual relations for specific schemata that are included in the subject matter. In general, close examination of the content of the schemata across the curriculum has not typically been an enterprise of the psychologist. Typically, the psychologist is happy to show that schemata are useful theoretical constructs and then proceed to explore general properties of their acquisition and use.

Having identified important schemata, mathematics educators should explore instructional activities that will help students to (a) learn the conceptual relations that constitute a schema, (b) learn how to translate verbal problems into that conceptual structure, and (c) learn to recognize *when* to use a particular schema. All three components are essential for successful problem solving with understanding. It may be necessary to help students focus their attention on these issues. One way might involve the use of exercises that ask students simply to describe the conceptual relations—no computations expected.

I could not locate any cognitive studies that examined the effects of teacher-led didactic or interactive instruction on algebra learning. This seems to be a place where mathematics educators could apply their instructional experience to mutual benefit. That is, if mathematics educators can produce documented examples of significant algebra learning, then cognitive psychologists should undertake the challenge of accounting for what has happened.

The following are other issues that are probably relevant to developing effective algebra instruction, though these issues have not been specifically studied for algebra.

In typical instructional settings, students often do not know what to focus on. For example, Benjamin Bloom, the Charles H. Swift Distinguished Service Professor Emeritus in Education at the University of Chicago, was a participant in an experiment in which university faculty attended specially prepared "typical" undergraduate physics lectures. After watching a laboratory demonstration, Bloom commented that he did not know what to pay attention to (Tobias, 1987). Another participant in the same experiment wrote:

> It seemed to me during these lectures that I lacked any framework of prior knowledge, experience or intuition that could have helped me order the information I was receiving. I had no way of telling what was important and what was not. I had difficulty distinguishing between what I was supposed to be learning and what was being communicated merely for purposes of illustration or analogy. Worse yet, as one of the commentators commented, I could not tell whether I understood or not. Nothing "cohered." (Tobias, 1986, p. 39)

Similar problems were observed when students and instructors were asked to underline the relevant material in a physics chapter (Dee-Lucas, 1984). Unlike the instructors, the students usually marked material that was irrelevant to the conceptual issues being presented in the chapter.

Even when students do know what to focus on, they often do not know what to do with the information, as observed in several studies that examined student ability to identify the relevant parts of a problem. Hayes and colleagues found that students could identify the relevant information in a logical puzzle even if they could not solve the problem (Hayes, Waterman, & Robinson, 1977; Robinson & Hayes, 1978). Similarly, Chi, Feltovich, and Glaser (1981) found that beginning physics problem solvers were capable of identifying the relevant parts of a physics problem with the same facility as expert problem solvers in that area. However, the beginners often did not know what to do with the information that they identified.

Thus, we can see that students need help focusing their attention on relevant relations and deciding what to do with those identified relations, a conclusion that is consistent with the cognitive analysis of problem solving.

Finally, the focus of the cognitive analysis of learning has been to understand the growth of effective problem-solving ability. However, having this ability, alone, may not mean that students have had a satisfactory education in algebra. There are many curricular issues that educators are concerned with that go well beyond the cognitive-psychological interest in effective problem solving. For example, algebra instruction is often viewed as an opportunity to help students acquire some insight into mathematical thinking. This goal of algebra instruction does not guarantee the discovery of effective methods for communicating schematic structures and problem-solving operators for algebra word problems. Nevertheless, by drawing on some of the insights provided by cognitive psychology regarding the nature of algebra learning and problem solving, mathematics educators will be able to develop more effective instruction to meet their educational goals.

REFERENCES

Bobrow, D. G. (1968). Natural-language input for a computer problem-solving system. In M. Minsky (Ed.), *Semantic information processing* (pp. 135-215). Cambridge, MA: MIT Press.

Bundy, A. (1975). *Analyzing mathematical proofs* (DAI Research Report No. 2). Edinburgh, Scotland: University of Edinburgh, Department of Artificial Intelligence.

Carry, L. R., Lewis, C., & Bernard, J. E. (1980). *Psychology of equation solving: An information processing study* (Technical Report) Austin: University of Texas.

Chi, M. T. H., Feltovich, P., & Glaser, R. (1981). Categorization and representation of physics problems by experts and novices. *Cognitive Science, 5*, 117-123.

Committee on Fundamental Research Relevant to Education. (1977). *Fundamental research and the process of education.* Washington, DC: National Academy of Sciences.

Cooper, G., & Sweller, J. (1987). Effects of schema acquisition and rule automation on mathematical problem-solving transfer. *Journal of Educational Psychology, 79*, 347-362.

Dee-Lucas, D. (1984, April). *Expert and novice strategies for processing scientific texts.* Paper presented at the annual meeting of the American Educational Research Association, New Orleans, LA.

Greeno, J. G. (1983). Forms of understanding in mathematical problem solving. In S. G. Paris, G. M. Olson, & H. W. Stevenson (Eds.), *Learning and motivation in the classroom* (pp. 83-111). Hillsdale, NJ: Lawrence Erlbaum.

Hayes, J. R., & Simon, H. A. (1974). Understanding written problem instructions. In L. W. Gregg (Ed.), *Knowledge and cognition* (pp. 167-200). Hillsdale, NJ: Lawrence Erlbaum.

Hayes, J. R., Waterman, D. A., & Robinson, C. S. (1977). Identifying relevant aspects of a problem text. *Cognitive Science, 1*, 297-313.

Hiebert, J. (Ed.). (1986). *Conceptual and procedural knowledge: The case of mathematics.* Hillsdale, NJ: Lawrence Erlbaum.

Hinsley, D. A., Hayes, J. R., & Simon, H. A. (1977). From words to equations: Meaning and representation in algebra word problems. In M. A. Just & P. Carpenter (Eds.), *Comprehension and cognition* (pp. 89-106). Hillsdale, NJ: Lawrence Erlbaum.

Husén, T. (1984). Issues and their background. In T. Husén & M. Kogan (Eds.), *Educational research and policy: How do they relate?* (pp. 1-36). New York: Pergamon Press.

Kintsch, W., & Greeno, J. G. (1985). Understanding and solving word arithmetic problems. *Psychological Review, 92*, 109-129.

Kotovsky, K., Hayes, J. R., & Simon, H. A. (1985). Why are some problems hard? Evidence from Tower of Hanoi. *Cognitive Psychology, 17*, 26-65.

Lewis, C. (1981). Skill in algebra. In J. R. Anderson (Ed.), *Cognitive skills and their acquisition* (pp. 85-110). Hillsdale, NJ: Lawrence Erlbaum.

Matz, M. (1983). Toward a computational theory of algebraic competence. *Journal of Mathematical Behavior, 3*, 93-166.

Mayer, R. E. (1981). Frequency norms and structural analysis of algebra story problems into families, categories, and templates. *Instructional Science, 10*, 135-175.

Mayer, R. E. (1982a). Different problem-solving strategies for algebra word and equation problems. *Journal of Experimental Psychology: Learning, Memory and Cognition, 8*, 448-462.

Mayer, R. E. (1982b). Memory for algebra story problems. *Journal of Educational Psychology, 74*, 199-216.

Mayer, R. E., Larkin, J. H., & Kadane, J. B. (1983). A cognitive analysis of mathematical problem-solving ability. In R. J. Sternberg (Ed.), *Advances in the psychology of human intelligence* (Vol. 2, pp. 231-273). Hillsdale, NJ: Lawrence Erlbaum.

National Council of Teachers of Mathematics. (1986, April). *Position statements.* Reston, VA: Author.

Paige, J. M., & Simon, H. A. (1966). Cognitive processes in solving algebra word problems. In B. Kleinmuntz (Ed.), *Problem solving: Research, method, and theory* (pp. 51-119). New York: Wiley.

Reed, S. K. (1984). Estimating answers to algebra word problems. *Journal of Experimental Psychology: Learning, Memory and Cognition, 10*, 778-790.

Reed, S. K. (1987). A structure-mapping model for word problems. *Journal of Experimental Psychology: Learning, Memory and Cognition, 13*, 124-139.

Reed, S. K., Dempster, A., & Ettinger, M. (1985). Usefulness of analogous solutions for solving algebra word problems. *Journal of Experimental Psychology: Learning, Memory and Cognition, 11*, 106-125.

Reed, S. K., & Evans, A. C. (1987). Learning functional relations: A theoretical and instructional analysis. *Journal of Experimental Psychology: General, 116*, 106-118.

Riley, M. S., Greeno, J. G., & Heller, J. I. (1983). Development of children's problem-solving ability in arithmetic. In H. P. Ginsburg (Ed.), *The development of mathematical thinking* (pp. 153-196). New York: Academic Press.

Robinson, C. S., & Hayes, J. R. (1978). Making inferences about relevance in understanding problems. In R. Revlin & R. E. Mayer (Eds.), *Human reasoning*. Washington, DC: Winston/Wiley.

Sleeman, D. H. (1984). An attempt to understand students' understanding of basic algebra. *Cognitive Science, 8*, 387-412.

Sweller, J., & Cooper, G. A. (1985). The use of worked examples as a substitute for problem solving in learning algebra. *Cognition and Instruction, 2*, 59-89.

Tobias, S. (1986, March/April). Peer perspectives on the teaching of science. *Change, 18*(2), 36-41.

Tobias, S. (1987, February). *Outsiders and insiders: Social factors that determine science avoidance and alienation*. Invited lecture presented at the Second National Science, Technology, and Society Conference, Washington, DC.

Van Dijk, T. A., & Kintsch, W. (1983). *Strategies of discourse comprehension*. New York: Academic Press.

Three Ways of Improving Cognitive Studies in Algebra
or
A Reaction to: "Cognitive Studies of Algebra Problem Solving and Learning"

Robert B. Davis

Curriculum Laboratory
University of Illinois at Urbana-Champaign

WHAT I HAD EXPECTED

Before I had thought much about the specific person who was delivering the paper on cognitive studies of algebra problem solving and learning, I had already prepared a list of concerns of my own. It seemed to me that there were three very common errors that might be made in a typical report on this topic and that would need to be corrected in the reaction. So I came prepared!

Of course, Seth Chaiklin made none of the errors that I came prepared to deal with. I want to deal with them nonetheless, because mathematics education researchers in general do make these errors, and they need to stop doing so.

Error Number One

First, I feared that the reported studies would focus on the usual kinds of "word problems" and would emphasize short-term (and shortsighted) strategies for increasing the odds that a student would seem to do the right thing, even though that student might have virtually no understanding of the problem. We have seen many studies of this type, and they probably tend to lead us in the wrong direction. In the first place, those typical word problems of the past are NOT effective ways to develop mathematical ideas in most students; the main difficulties in these problems usually come in the implicit (and unreasonable) assumptions.

Rather than risk insulting any friends, let me make up a truly bad word problem for which I alone am responsible:

> Bill is mowing his front lawn, which is rectangular in shape. His lawn mower cuts a swath 2 feet wide. His lawn is 16 feet wide. How many passes with his lawn mower must Bill make?

This problem contains at least one difficulty in its underlying assumptions: If, as is likely, we are supposed to ignore the fact that good lawn mowing

115

usually requires overlapping swaths, we should answer, "Eight passes." But if we do this, are we learning mathematics as "a reasonable response to a reasonable challenge"? Are we learning mathematics as "a generally good description of reality"? Or, are we learning a rather formalized ritual of academic or "bookish" nonsense? The available evidence, of which there is a great deal, indicates that for most students this sort of problem solving, continued day after day after day, will seem unrealistic, foolish, and irrelevant.

Of course, this problem could be salvaged. One could ask students, "What can you say about the number of passes that will be required?"—to which it is entirely sensible to answer that it must be at least 8 and will probably be somewhat more than that.

I have a reason for emphasizing this matter. In the important curriculum revision work of recent decades, one of the goals—and, I would argue, an important one—was to get rid of artificial word problems and replace them with something more genuine. The reason for this was the desire to give students a more positive view of mathematics, to show that it really *is* a sensible subject. (In the British film *I Do, and I Understand*, produced by the Nuffield Mathematics Project, the distinction between a sensible problem and an irrelevant bookish-type problem is described in the words, "a *real* problem, and not what we've taken to be a problem in the past, that was merely a sum wrapped 'round in words.")

Why is this relevant to a research agenda? Because the relation between research and practice is a complicated one, indeed. If we carry out research studies on certain kinds of problems that are commonly found in today's textbooks, the mere fact of these studies' being undertaken helps legitimize those practices. Those of us who attach great importance to changing common practices and commonly used textbooks cannot honestly welcome studies that seem to accept the value of both. We should, instead, study how students deal with more genuine (and often more complex) problems— problems that have the important redeeming virtue of "making sense."

If one does make the error of focusing on highly artificial word problems, one can compound the error by accepting very short-term strategies for producing what can look like correct student performance. Rules for "looking as if you know what you're doing, even though you don't, really" abound in all elementary mathematics courses, from arithmetic through calculus. In arithmetic we have rules such as, "If it says 'altogether,' then add." In calculus we have rules such as, "When in doubt, differentiate." Most readers will have encountered many of these. Such rules do NOT result in student understanding, nor do they really even pretend to do so—but they do sometimes result in better performance on certain kinds of tests. Of course, in an oral examination situation, this kind of fraud is easily detected through the use of follow-up questions.

Why does research entail the risk of looking at the wrong kinds of tasks and goals? One reason is that research responsibilities are often parceled out among different specialists, and when research studies are undertaken by people whose main interest is not the teaching and learning of mathematics, the researchers may simply take common present-day practice as an indication of what *ought* to be done. They may believe that the goal is merely to find ways to do it more efficiently. I would argue that inefficiency is not the major difficulty that we face at the present time. Our biggest problem is that *the wrong kind of learning experiences are being aggregated into the wrong kind of courses, aimed at producing the wrong kind of mathematical "knowledge."* A more fundamental kind of research is needed if we are to make significant progress.

Error Number Two

If present mathematics courses are not leading in an optimal direction, what sort of course would constitute an improvement? Well, what different kinds of courses are possible?

Let me start by describing two extremes. On the one hand, we might have a course that treats mathematics as a collection of highly specific skills and techniques. "Collecting like terms" is usually on the list. So is "removing parentheses." So are "multiplying out" and "solving linear equations." "Solving quadratic equations" is on the list, but "solving cubic equations" is not. The kind of course I am trying to describe approaches these topics as individual small skills, presented to students as rituals to be practiced until they can be executed in the proper, orthodox fashion.

I am looking at such a presentation right now, which I shall not identify. It begins by asking the student to simplify the expression:

$$3(x + 4) - 2x + 5.$$

No motivation has preceded this. It is somehow assumed that the student will *want* to simplify this expression. It is further assumed that the student will understand what is meant by "simplifying." But, which of these expressions is simpler:

$$\frac{1}{\sqrt{2}} \quad \text{or} \quad \frac{\sqrt{2}}{2} \, ?$$

Does it not depend upon what you plan to do next with the result?

This approach to algebra reminds me of the way I was taught handwriting—we practiced circular movements and made O's, we practiced looping motions and made *l*'s, and so on. I did learn to write, but this process never struck me as sensible. I never really knew what I was supposed to be doing.

From my perspective, these courses have the following structure: The student is asked to perform some fragmentary piece of some ritual. The student sees no purpose or goal to this activity, other than extrinsic goals (such as pleasing the teacher) or competitive goals (such as doing it better than Joey does). Consequently, the student sees no reason why the ritual is performed in one way and not another.

The theory underlying such courses seems to be: *If the students spend enough time practicing dull, meaningless, incomprehensible little rituals, sooner or later something WONDERFUL will happen.* I have never shared this optimism.

At the opposite extreme, we have a course that begins with some task that students find interesting. Warwick Sawyer's game of "Guessing Functions" (or "Guess My Rule") usually works well. Some "magical tricks" with numbers catch the attention of some students. Stein and Crabill (1972) make excellent use of the task of maximizing the volume of a certain tray that is made by folding up flaps on a piece of paper. My own project uses the task of making up an open sentence that will be true no matter what numbers are written in a □.

Is a course based on problem solving really different from a skill-oriented course? One professional colleague (who is in fact the author of the "simplify $3(x + 4) - 2x + 5$" lesson) argues that it is not. In both cases (he says) we hope students will accept the legitimacy of some task and will therefore carry out the procedure we teach them—indeed, will practice it until they are good at it.

I disagree. In the skill lesson, we are dealing with a ritual—since we do not know the goals, we cannot creatively devise our own strategies. Since we have no context, we cannot analyze the mathematical situation. What are we left with? We can imitate this little piece of ritual, and that is all we can do.

I do not recognize this as mathematics. *Mathematics*, to me, means being faced with some interesting challenge, being confused as to how to formulate a more precise version of that challenge, trying to decide what an answer might look like, looking for clues and cues and suggestions and patterns, trying to interpret these cues and clues, trying to devise a strategy for solving the problem, working to implement that strategy, and deciding how satisfied we are with the result. Nor is it always that simple—we do not necessarily go through that sequence just once. On the contrary, we may have to go through parts of the sequence many times.

Can young people deal with this level of complexity? Of course—they do it all the time! Suppose it has snowed. Some youngsters want to do something interesting with all that wonderful snow! They decide to build a fort. They find there is one vulnerable corner to their fort. They find ways to make it more secure. Run this sequence through your imagination a few times, and you can easily visualize those youngsters working through all of the processes I listed above.

Research cannot look at problem solving in a vacuum—student activity will occur within, or in relation to, some kind of course. Which kind of course? I would argue that observing both kinds of courses, getting good descriptions of them, and getting good descriptions of how students differ depending upon which kind of experience they have had, ought to be a major part of the research agenda. To be effective, this work would have to be cast in a "cognitive" format, and it would have to focus on problem solving. Hence it is part of the answer to what cognitive studies of algebra problem solving and learning ought to be like.

Studies that accept the presently commonplace skills-and-rituals course seem to point us in the wrong direction. Studies that compare and contrast the rituals course versus the problem-based course would seem to point in a very useful direction, indeed.

Error Number Three

Many—really, most—studies focus on what a student writes and largely ignore what that student thinks. Yet what the student thinks is far more fundamental than what the student writes. The tradition of educational studies of mathematical behavior has been to ignore human thought processes. Most of us have heard the argument that, because we can never know what goes on in the human mind, it is unscientific to speculate on such matters.

No real science has ever given itself such narrow constraints. Imagine saying, "We'll never know what the earth was like before humans started to keep records, and therefore it would be unscientific to speculate on such matters." Neither geology nor anthropology nor bio-chemistry accepts such restrictions—they do think about such matters and do so in a profitable way. Similar examples can easily be found in physics, chemistry, astronomy, biology, and medicine. We cannot afford to rule out the serious analysis of how humans think about mathematics. Some discussion of this question is found in a later chapter, "Research Studies in How Humans Think About Algebra" (Davis, this volume).

WHAT I FOUND

Of course, I prepared my remarks before I knew who would be reporting on cognitive studies of algebra problem solving and before I had an opportunity to learn more about Seth Chaiklin's approach. It is now clear that none of my concerns pertain to Seth's paper, but I still want to insert them into the record, because the issues I have raised are important ones in general. We need to encourage research studies that focus on what matters, rather than studies that legitimize practices that need to be changed.

REFERENCE

Stein, S. K., & Crabill, C. D. (1972). *Elementary algebra—A guided inquiry*. Boston, MA: Houghton Mifflin.

Robust Performance in Algebra: The Role of the Problem Representation

Jill H. Larkin

Department of Psychology
Carnegie Mellon University

The mathematical structure and algorithms of algebra are simple and elegant. Nonetheless, many people seem never to grasp algebraic techniques reliably and, despite repeated instruction, continue to perform unreliably. The reasons for this difficulty are a puzzle and a frustration to psychologists, mathematics educators, and teachers of all fields that build on mathematics. The purpose of this paper is to present a cognitive model of algebraic reasoning that I believe may explain at least in part these resilient difficulties. In developing the algebra model, this paper discusses general cognitive science techniques for modeling complex human behavior. It also proposes an experimental test of the model by using it to design instruction.

This paper addresses only the symbol system of algebra—understanding and manipulating expressions and equations that include symbols for variables. It does not discuss (except in a few comments) all the questions that arise in relating this formal system to problems described in words.

The first section applies cognitive modeling techniques to develop a simple model of competent performance in algebra. The next section considers typical errors made by people who do not perform competently and argues that a model which can produce such errors must be radically different in important ways from the competent model. A final section uses these models to propose a possible new method for teaching algebra.

A SIMPLE MODEL OF COMPETENT PERFORMANCE IN ALGEBRA

The central technique of cognitive science is modeling problem-solving behavior in the following way: A problem is considered as a *data structure* that includes whatever information is available about the problem. We then ask what kind of *program* could add information to that data structure to produce a solution to the problem. Because we want to have a model that explains human performance, we require that the model add information to the data structure in orders consistent with the orders in which humans are observed to add information. Additionally, the solutions produced should be similar to those that humans produce.

This approach is sometimes misunderstood as a statement that human beings are in some sense "like" computers. I do not believe that this is true. Some years ago psychologists developed interesting models of simple learning by modeling human learning with matrix operators (Markov models). No one would claim that humans are like matrices. Writing a model in LISP (a computer language commonly used in cognitive science) is different from writing a model in matrices, only in that LISP provides more powerful mathematical modeling tools. In either case (matrices or LISP) a computer is useful in implementing and checking the predictions of the model. But, in neither case is it useful to argue that humans are like matrices or like a LISP program.

It is also common to infer that building computer models to help understand learning and problem solving somehow implies that students should be taught computer-like rules or algorithms. Again, this is definitely not the case. If I believed that humans could adequately understand and use a simple rule-based approach (Sowder, 1980), I would just write down the rules, as do the advocates of this approach. It is *only* because I am convinced that human thoughts interact in complex and unpredictable ways that I find the richness of a computer language a useful tool in characterizing human thinking and learning.

With these views in mind, let us ask how we can model competent performance in algebra with a computer program. Our model starts with a data structure that captures the information initially present in a problem. Let us consider a problem like the following, a moderately complex linear equation in one variable:

$$2x - 3(5x - 7) = 4 + 9x$$

A model for solving equations like this will consist of a data structure that can represent equations and a program that can act on this data structure to solve the equations correctly with the same kinds and sequences of steps that competent human solvers use.

The simplest data structure that captures the information in the preceding equation is just an ordered string in which each of the algebraic symbols above follows its predecessor. In order to solve this equation, a model solver must be able to add to this string data structure the following information:

- It must be able to identify the various types of symbols—including numbers, variables, operators, equal sign, and parentheses— because later the program must have separate procedures indicating what to do with different kinds of symbols.

- It must be able to discover and encode the equation structure dictated by operator precedence and by the parentheses, so that later the program will be able to move and combine symbols appropriately.

There are two ways to build a model that adds this needed information. The model might work with the given string data structure. Then every algebraic inference must contain explicit tests to identify different kinds of symbols and their relation to each other. Alternatively, one can divide the problem-solving program into two parts. A *representation-building* part acts on the given string data structure to build a richer data structure, which is then used by the central *problem-solving* part of the program. The problem-solving part of the program can be simpler and easier to comprehend if the representation-building part develops a good data structure. This division of labor reflects the distinction between "understanding" and "solving" made by Hayes and Simon (1976).

Besides, it seems likely that skilled human solvers build a rich problem representation before beginning to solve a problem. In algebra, a competent solver can, without solving an equation, easily identify the different kinds of symbols and can group together substrings determined by operator precedence. I suggest that these things are automatically and easily "seen" by a competent solver whenever he or she looks at an equation and that these features remain available in the mind throughout the solution process.

Figure 1 shows an equation written in standard form and a diagram of a data structure for this equation. This data structure includes explicitly the following information, which will be used by the problem-solving program:

- All symbols have been tagged with an indication of their type;
- The implicit multiplications have been made explicit with * operators, except in the simple case of a number times a variable (e.g., $2x$), where the multiplication has been left implicit;
- Finally, and most importantly, the original linear string has been reorganized into a tree with connections explicitly indicating the order in which the operations must be performed.

This tree data structure makes explicit the hierarchy among the operators in the equation. Branches joined by operators just below the equal sign correspond to parts of the equation that can be "moved" from one side of the equation to the other by applying the same operator to both sides of the equation. It is illegal to move numbers or variables deeper in the tree.

There is some arbitrariness in what one puts in the data structure. Whatever one leaves unrepresented there must then be handled by the problem-solving part of the program. For example, in Figure 1, nodes involving subtraction are ordered, as indicated by the arrow appearing at each subtraction node. The reason is that the subtraction operator must be applied to its two arguments in a particular order. Division, of course, has the same property. In contrast, the multiplication and addition nodes are unordered (have no arrows), reflecting the commutativity of these operations. The unordered nodes allow the rules that operate on this data structure to perform these operations in whatever order is convenient. Thus, the data structure in Figure 1 shows explicitly the different commutative properties of addition and subtraction. It would also be possible to use a data structure

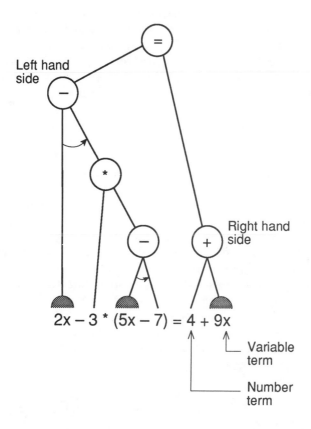

Figure 1. Hierarchical data structure for the algebraic equation $2x - 3(5x - 7) = 4 + 9x$.

that omits the ordered nodes for subtraction. Then the problem-solving part of the program, instead of just being allowed to interchange any unordered node, would have to contain tests that allowed it to interchange some nodes (+ and *) but not others (− and /).

In general, I find it most informative to include in the data structure those things a human solver can "see" immediately when looking at a physical display (Larkin & Simon, 1987). That is, the data structure should always contain elements corresponding to the physical objects or primitive paper marks in the display (here, the numbers, variables, and other symbols). Additionally, however, I design the data structure to contain information a human solver could report immediately and easily in just looking at the display. Here, the data structure captures such features as the grouping of the equation according to operator precedence, the meaning of the implicit multiplication, and the commutativity properties. In fact, when typesetters set equations, they use variable spacing to reflect these groupings (*Words into Type*, 1974), so that a printed equation may actually have some of this information represented explicitly. The reason for encoding this information in the data structure is that I want the computer model to have easily available at all times the background information that the human solver brings to the problem—that is, the information that the human solver can essentially read off the display.

Although we are currently considering expert solvers, this is perhaps the place to note that novice solvers almost certainly do not have available all the information that experts can so readily "see." In fact, I will argue that, in the case of algebra, this may be a major source of common difficulties in solving equations.

What kind of program could start with a simple string and produce the annotated data structure in Figure 1? A conceptually simple (and often practical) way of building such a program is to construct it as a set of inference rules, for example, like those shown in Table 1. Again, remember that these are not the kind of rules humans know or state explicitly or that would be taught to students. Table 1 is a listing, in English, of rules that appear in a computer program that starts with a string and builds meaningfully annotated data structures for linear algebraic equations. The inference rules in parts (a) and (b) of Table 1 (called production rules) form the representation-building part of our algebra program. If the program exe-

Table 1
Production Rules for Solving Linear Equations Having Integer Coefficients and Written as Strings: (a) Identifying the Type of Each Symbol, (b) Building a Labeled Tree Representation, and (c) Solving the Equation

---

(a) *Productions for identifying symbols*

1a. IF a symbol is a member of the set {0,1,2,3,4,5,6,7,8,9} or is composed of an unbroken substring of such symbols,
   THEN it is a number.

1b. IF a symbol is a member of the set {a,b,c, . . . ,z},
   THEN it is a variable.

1c. IF an unbroken substring of symbols consists of a number directly followed by a variable,
   THEN it is a variable term; the number is the coefficient of the variable.

2a. IF a symbol is a member of the set {),], } },
   THEN it is a right parenthesis (<rp>).

2b. IF a symbol is a member of the set { (, [, { },
   THEN it is a left parenthesis (<lp>).

3a. IF a symbol is a member of the set { = },
   THEN it is an equal sign (<eq>).

3b. IF a symbol is a member of the set { +, −, *, / } and follows a symbol other than <lp> or <eq>,
   THEN it is an operator (<op>).

4.   IF a variable or <lp> is not immediately preceded by a number,
   THEN insert the symbol 1 in front of it.

5.   IF a symbol is a member of the set { +, − } and is either the first symbol in the string or immediately follows <lp> or <eq>,
   THEN the symbol is part of the number that follows it.

Table 1 (continued)

(b) *Productions for building structure*

6.   IF   an unbroken substring of symbols is either a number or a variable term,
    THEN it is a generalized number (GN).

7a.   IF   there is a sequence of the form: GN <op> GN,
    THEN make pointers from the operator to the adjacent GNs and thereafter
            consider this structure a new GN.

7b.   IF   an operator with pointers is a member of { $-$, / },
    THEN make a directional arrow from the first pointer to the second.

8a.   IF   there is a sequence of the form: GN <op1> GN <op2> GN and <op1>
            is a member of { *, / } and <op2> is a member of { $+$, $-$ },
    THEN GN <op1> GN is a new GN.

8b.   IF   there is a sequence of the form: GN <op1> GN <op2> GN and <op2>
            is a member of { *, / } and <op1> is a member of { $+$, $-$ },
    THEN GN <op2> GN is a new GN.

9.   IF   there is a matched pair of parentheses,
    THEN the parentheses and the symbols between them are a GN.

10.   IF   there is a sequence of the form: GN1 GN2 and one of the GNs is neither
            a number nor a variable,
    THEN insert the symbol * between the two GNs.

11.   IF   there is a sequence of the form: GN1 <eq> GN2,
    THEN make pointers from the equal sign to the adjacent GNs; label GN1 the
            "left hand side" of the equation and GN2 the "right hand side" of the
            equation.

(c) *Productions for solving the equation*

12a.   IF   the top operator on the left hand side of the equation is a member of
            { $+$, $-$ } and a pointer from this operator points to a number,
    THEN delete the number and the operator preceding it from the left hand side
            and place it at the end of the right hand side, preceded by the "opposite"
            operator.

12b.   IF   the top operator on the right hand side of the equation is a member of
            { $+$, $-$ } and a pointer from this operator points to a variable term,
    THEN delete the variable term and the operator preceding it from the right hand
            side and place it at the end of the left hand side, preceded by the
            "opposite" operator.

13a.   IF   the two pointers from an operator both point to numbers,
    THEN perform the indicated operation.

13b.   IF   the two pointers from a $+$ or $-$ operator both point to variable terms,
    THEN combine the terms by adding or subtracting the coefficients.

13c.   IF   the two pointers from a * operator point to a number and a variable term,
    THEN multiply the number by the coefficient of the variable term.

14.   IF   there is a sequence of the form: <lp> GN <rp>,
    THEN remove the parentheses by applying the distributive law.

15.   IF   there is a sequence of the form: GN1 <op1> GN2 <op2> GN3 and
            <op1> and <op2> are both members of the set { $+$, $-$ } and neither
            GN1 <op1> GN2 nor GN2 <op2> GN3 can be combined,
    THEN reorder the sequence as: GN1 <op2> GN3 <op1> GN2.

16.   IF   the left hand side of the equation consists of a single variable term and
            the right hand side consists of a single number,
    THEN the solution of the equation is the number on the right hand side divided
            by the coefficient of the variable.

---

cutes all of these rules until none of them applies, then it produces the representation shown in Figure 1.

To illustrate, the computer program listed in Table 1 starts with a string like the following:

$$2x - 3(5x - 7) = 4 + 9x$$

The program "reads" from left to right across the equation. When it reads the symbol 2, Rule 1a executes, labeling this symbol as a number. Then the program reads the next symbol, labeling it as a variable, and the next, labeling it as an operator. This process continues until all symbols are labeled. Notice that Rule 4 inserts the symbol 1 whenever there is a "missing" coefficient, and Rule 5 distinguishes the + and − used for positive and negative integers from the addition and subtraction operators. These are examples of making explicit, in the program, features of the equation structure that a human solver "knows" just by looking at the equation.

Once the basic symbols are identified, the program begins building structure, matching the IF parts of Rules 6-11 to the current state of the equation. For example, Rule 6 labels the symbol 3 as a GN, Rule 9 labels $(5x - 7)$ as a GN, and then Rule 10 inserts the symbol * between them.

Notice that this program does not work in the usual way of structured languages like FORTRAN and PASCAL. There is no order to the rules. Each rule simply executes when its conditions are satisfied. This flexibility is a major reason why production systems are a useful tool in understanding human reasoning. The variable order also distinguishes this program from any ordered algorithm or "rule list" one might try to teach a student.

After the symbols are labeled and the hierarchical data structure is constructed, the problem-solving part of the program takes over, as shown in Table 1(c). Rules 12a and 12b move numerical terms to the right hand side of the equation and variable terms to the left. The next three rules combine all combinable sequences. Rule 14 "clears the equation of parentheses," using the distributive law, and Rule 15 reorders terms so they can be combined. The final rule solves the simplified equation for the value of the unknown.

As an example, let us see how the problem-solving part of the program works with the data structure in Figure 2(a) for our model equation:

$$2x - 3*(5x - 7) = 4 + 9x$$

At the beginning, two of the productions (12b and 14) from Table 1(c) apply. The program can begin with either step. (Here and elsewhere I am ignoring some subtleties of exactly how production systems resolve the conflict when two or more rules apply—see Brownston, Farrell, Kant, & Martin, 1985, for a more complete discussion of production systems.) Suppose the program begins by moving $9x$ to the left hand side of the equation, that is, by subtracting $9x$ from both sides. This production produces the new hierarchical tree shown in Figure 2(b). Two more productions (15 and 13b) execute to combine the $2x$ and $9x$ to get $-7x$ as in Figure 2(c). Then the program clears parentheses, as shown in Figure 2(d). Table 2 shows one complete sequence of steps that could be followed in solving the model equation, together with the production applied at each step.

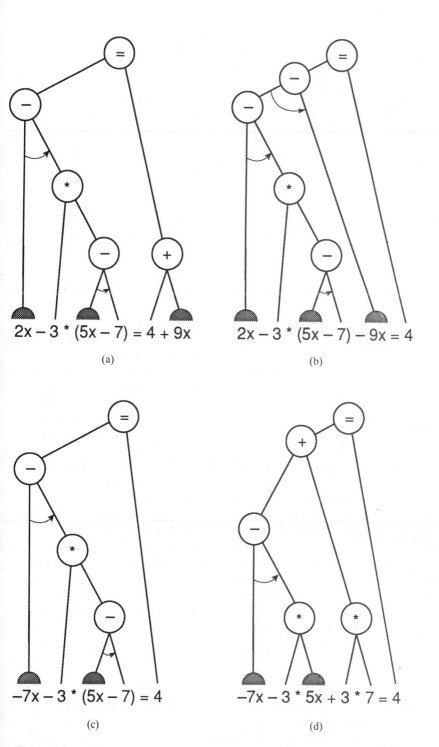

$$2x - 3 * (5x - 7) = 4 + 9x$$

(a)

$$2x - 3 * (5x - 7) - 9x = 4$$

(b)

$$-7x - 3 * (5x - 7) = 4$$

(c)

$$-7x - 3 * 5x + 3 * 7 = 4$$

(d)

Figure 2. Changes in the equation data structure during the first few steps of the computer solution.

Table 2
Steps of an Algebraic Solution Done by the Computer Model, Using the Production Rules in Table 1(c)

Equation:	$2x - 3*(5x - 7) = 4 + 9x$	
Step 1:	$2x - 3*(5x - 7) - 9x = 4$	Rule 12b
Step 2:	$2x - 9x - 3*(5x - 7) = 4$	Rule 15
Step 3:	$-7x - 3*(5x - 7) = 4$	Rule 13b
Step 4:	$-7x - 3*5x + 3*7 = 4$	Rule 14
Step 5:	$-7x - 15x + 3*7 = 4$	Rule 13c
Step 6:	$-7x - 15x + 21 = 4$	Rule 13a
Step 7:	$-22x + 21 = 4$	Rule 13b
Step 8:	$-22x = 4 - 21$	Rule 12a
Step 9:	$-22x = -17$	Rule 13a
Step 10:	$x = 17/22$	Rule 16

COMMON ERRORS AND A MODEL THAT CAN PRODUCE THEM

We have seen how a relatively simple model, consisting of a data structure (a tree) and a set of rules that operate on it, can produce competent, human-like solutions to a simple type of algebraic equation. Of course, we would have to add more rules in order for the system to handle more complex equations smoothly. But I believe that the basic structure and design are capable of supporting a more elaborate production system. Indeed, the model of Bundy and his colleagues (1979) works almost exactly this way to solve essentially any algebraic equation in one unknown (including quadratics).

Now let us turn to the problem of students who resiliently fail to learn algebra. On the basis of our model, the difficulty could be attributed to the rules, to the data structure, or to both. As I shall discuss later, where the difficulty lies has profound implications for what to do about it. Most people addressing this problem (Carry, Lewis, & Bernard, 1980; Sleeman, 1982)—including, I believe, educators—have attributed the problem to the rules and thus have worked to develop rules that produce the kinds of errors poor algebra students make.

Let us therefore start by again using the hierarchical data structure in Figure 1 and asking what rules would be required to produce just a few sample novice errors like the following:

Error 1. Move an inaccessible term across the equal sign
$$2x - 3(5x + 7) = 4 + 9x$$
$$2x - 3(5x) \qquad = -3 + 9x.$$

Error 2. Remove parentheses without multiplying the second term
$$2x - 3(5x + 7) = 4 + 9x$$
$$2x - 15x + 7 \quad = 4 + 9x.$$

In contrast to these common errors, there are conceivable errors that have never appeared in any of the catalogues of actual student errors, for example:

Error 3. Remove parentheses without multiplying the first term
$$2x - 3(5x + 7) = 4 + 9x$$
$$2x - 5x - 21\ \ = 4 + 9x.$$

What inference rules can we write that will produce the first two errors but not the third? Recall that the problem data structure (Figure 1) used by our competent program made no distinction between the two arguments pointed to by a + operator. Therefore, any program based on this data structure should make both Errors 2 and 3. The reason is that the addition operator points symmetrically to both of its arguments. Therefore, any program that produces an error involving one argument (Error 2) must also produce a symmetric error for the other argument (Error 3). Furthermore, it is hard to see how to devise a program that uses the data structure in Figure 1 and which will make Error 1—the 7 being moved across the equal sign is not at the top of the hierarchy. To produce this error, we would have to make it legal to move anything across the equal sign.

Suddenly, in contrast to the simple model we formulated for competent performance, we find that to explain novice behavior we must postulate new and very strange rules. Our model is beginning to seem like a very implausible characterization of student behavior. Let us therefore explore another way of building a model of novice performance: Suppose we do not keep the tree structure that supported the competent model but, instead, assume that novices may form a rather different data structure for equations. Certainly the simplest such structure is just the string of characters in the equation itself. Can we formulate a set of rules based on such a representation that will produce the kinds of errors to which students are prone? Consider the following "novice" rules:

N-Rule 1.   IF a term is connected to other terms by a + or − operator, THEN move it across the equal sign and change its sign.

N-Rule 2.   IF two symbols are separated by a left parenthesis, THEN ignore the parenthesis.

These simple rules, applied to a string data structure, produce exactly Errors 1 and 2 above, but not Error 3.

This analysis suggests that people who have continuing trouble learning algebra may look at an equation and "see" only the string of symbols, rather than the connected structure represented by the expert's tree. If this is the case, we can also explain why these people find it so difficult to learn correct rules. Consider, for example, in Table 1 the inference rule that uses the distributive law (Rule 14). This rule is invoked to break apart expressions whose terms cannot be combined directly. For example, in $2x - 3*(5x - 7)$ the terms in $5x - 7$ cannot be combined, so the expression is taken apart and combined with the 3, term by term, to form a new expression, which can then be combined with the $2x$.

But if a novice solver sees no structure in the string

$$2x - 3*(5x - 7),$$

then there is no reason to want to dissociate the $5x - 7$. Furthermore, if the 3 can simply be combined with the $5x$, the problem is just that much easier. Consider also the expert rule that only elements at the top of the tree may be moved across the equal sign. In a string there is no "top" and no sensible way of distinguishing movable from nonmovable elements. In short, for a student working with a string representation, expert rules simply do not make sense.

If the novice sees the substring enclosed in parentheses as part of a string, then it makes no sense to treat the symbols within the parentheses any differently from those outside. The significance of the parentheses as a grouping device is not meaningful in the novice representation. The rather odd rule about multiplying terms, starting at one parenthesis and continuing until you come to another, seems quite arbitrary. For the novice, the terms within the parentheses are no different from those outside. There is no reason to notice the parentheses or to remember to give them any special treatment.

I believe this view may help to explain why some people simply do not learn algebra, despite being informed of the correct rules and practicing them many times. The fact may be that, for these people, our expert rules simply make no sense. They try to use them on their string data structure, but the string structure provides no cues for distinguishing between legal and illegal combinations. Therefore, within a short time, if the learner forgets exactly how the rule worked, there is no basis for reconstructing it— no reason to think, "It makes sense to do it this way." On the contrary, the novice is quite likely to begin re-using the old, incorrect rules because they seem to make sense when applied to a string data structure.

A PROPOSED EXPERIMENT

I have proposed a strong hypothesis about the nature of at least some learners' difficulties in applying algebra with understanding and reliability. I have suggested that a big part of what we might mean by "understanding" an equation is being able to construct a good internal representation for it. I have demonstrated that in computer programs, a hierarchical representation works nicely in a program that solves equations correctly, whereas a string representation makes correct behavior difficult and incorrect behavior easy.

So far, though, this is just talk and computer programs. How can we test my hypothesis? Equivalently, how can we teach representation-constructing skills? I have found this to be a difficult question. If we want to teach skills, we can demonstrate procedures, ask learners to imitate us, and correct observed behavior. But a problem representation is largely internal. We cannot ordinarily make this process visible for students or observe their processes well enough to correct them.

One possibility is to try to make the representation visible through some display. For example, we used the display in Figure 1 to show the representation of the computer model. But the display in Figure 1 seems to me unsatisfactory as a representation of an internal data structure. Knowing that some displays are unsatisfactory, I have composed the following list of suggested criteria for good displays:

1. The visual display should be isomorphic to the representation we are trying to teach. If this representation is part of a computer model, then this isomorphism can be explicitly checked.

2. The most salient and problematic features of the internal data structure should be isomorphic to the most salient features of the visual display. Correspondingly, the visual display should not contain obvious features that misleadingly suggest relations that are not present in the correct data structure.

3. The display should use the notation usually used in solving problems of this kind. The goal is to help the learner solve problems independently and efficiently—generally from the usual representation. We do not want to produce long-term dependence on an arcane display without being able to transfer the new understanding to the usual display.

4. With the visual display, ordinarily difficult processes should be represented by easy perceptual processes. I have argued elsewhere (Larkin & Simon, 1987) that a picture or diagram can be worth 10,000 words exactly when, by using the diagram, one can substitute easy perceptual processes for complex and difficult inferencing.

5. If possible, the display should be memorable—easy to reconstruct mentally when solving problems in the future.

The preceding criteria are useful guidelines for recognizing a good display. I have found them less helpful in generating good displays. In fact, only after months of thinking of bad displays, which did not meet the preceding criteria, did I finally arrive at what I think is a good display for teaching the representation of equations.

Figure 3 shows an example of this display. Each of the general numbers from our computer model (see Table 1) is matched onto one tile in the display. Using this display, the rules of algebraic manipulation have simple visual counterparts—a tile may be moved across the equal sign only if it is at the bottom level. Only tiles of the same level can be combined and then only if they both reside on the same subtile or are separate tiles. When tiles are combined, each resulting term is written on a separate tile at the next lower level.

These rules do not, of course, comprise the whole of algebraic manipulations, but they are easy, visually simple rules that forbid exactly the kinds of mistakes that students commonly make. In this display, the constraints of the expert representation are obvious. For example, it just does not look reasonable to take a tile from the top and slide it across the equal sign. It does not look reasonable to take tiles from two different levels and combine them. But it does look reasonable that combinations must involve all of the contents of a single subtile. In short, for the very rules that students find the most confusing, the display replaces in-the-head inferences with intuitively logical, visual processing.

Having invented this representation, however, I still have only armchair suppositions. I have proposed that the lack of a good problem representation is the central difficulty for many students who are incompetent in algebra, and I have developed a visual display that is isomorphic to a good algebraic representation and which substitutes easy perceptual processes for difficult inferential processes.

How could these ideas be tested experimentally? (I will leave to teachers and mathematics educators far more expert than I to think about how such a display might be helpful in algebra instruction.) I will take on the more limited question—can I demonstrate (a) that using this display will enable previously incompetent algebra students to solve equations more effectively and (b) that this change is due to a change in internal representation and to other instruction (e.g., instruction in algebraic procedures)?

What I plan to do is to produce a simple computer instructional program with three sections:

- Solve equations placed on tiles;
- Given an equation, construct an appropriate set of tiles;
- Solve equations in conventional notation (with the option of placing them on tiles).

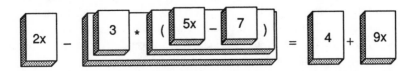

Figure 3. A display, reflecting equation structure, that students can manipulate in solving equations.

The emphasis of these programs will be on the display—other aspects of the program will be as simple as possible. As a start, I will probably simply judge each algebraic step right or wrong and give the student this feedback at each step.

The outcome might be something like the following: Prior to using the program, (a) students are unreliable in solving equations and simplifying expressions and (b) verbal interviews asking them to describe equations will suggest a string representation. After working with the program, I would hope to see evidence of (a) improved algebraic performance and (b) use of a hierarchical representation.

It would also be interesting to retest and interview students 1-3 months later. I have suggested that students may initially learn correct algebra rules but then forget them because the rules make no sense when applied to a string representation, and the string provides no constraints to prevent errors. If my conjecture is true, then a major strength of instruction leading to a better representation should be better retention of algebraic skills.

Such results would replicate some results of cognitive and developmental psychology. The Tower of Hanoi puzzle (used in many, many experiments) requires remembering that rings on a peg must always have smaller rings on top of larger rings. Modifications of this puzzle are easier for adults if stated in terms of acrobats, with the largest on the bottom and the smallest on top (Hayes & Simon, 1977; Kotovsky, Hayes, & Simon, 1985); the puzzle is harder if it is restated with smaller acrobats on the bottom. Similarly, in Klahr's experiments (Klahr & Siegler, 1978) even very young children have been able to solve parts of this puzzle if they work with empty painted cans stacked upside down on top of large pegs, so that it is impossible to stack the cans in incorrect sequences. In both of these adaptations of the puzzle, one rule of the problem is made apparent by the visible data structure, that is, by the display.

SUMMARY AND CONCLUSION

I have used the modeling techniques of cognitive science to try to elucidate a problem in algebra education that has always concerned me: Why are the simple rules of algebraic manipulation so difficult for some people to learn? I have argued that the difficulty may lie, not in the rules that are learned, but in the way the learner internally represents an algebraic equation. If a learner looks at an equation and sees an unstructured string, then many of the rules of algebra act on apparently arbitrary parts of the string. In this case, it is hardly surprising that these rules are very easy to forget or to misapply. In contrast, a competent solver can look at the same equation and automatically "see" a hierarchical structure. In this case, the rules of algebraic manipulation are well-defined as applying to certain forms in the structure.

Whether or not these conjectures are correct in the case of algebra, this description illustrates the importance of viewing problem solving as a program that acts on a data structure and noting that this program can be easier and simpler if the data structure is well-formed. Finally, I have proposed a visual display that might be used as part of a computer program to teach an effective representation for algebra.

REFERENCES

Brownston, L., Farrell, R., Kant, E., & Martin, N. (1985). *Programming expert systems in OPS5*. Reading, MA: Addison-Wesley.

Bundy, A., Byrd, L., Luger, G., Mellish, C., & Palmer, M. (1979). Solving mechanics problems using meta-level inference. *Proceedings of the Sixth International Joint Conference on Artificial Intelligence* (Vol. 2, pp. 1017-1027). Los Altos, CA: William Kaufmann.

Carry, L. R., Lewis, C., & Bernard, J. E. (1980). *Psychology of equation solving: An information processing study* (Final Technical Report). Austin: University of Texas, Department of Curriculum and Instruction.

Hayes, J. R., & Simon, H. A. (1976). The understanding process: Problem isomorphs. *Cognitive Psychology*, *8*, 165-190.

Hayes, J. R., & Simon, H. A. (1977). Psychological differences among problem isomorphs. In N. J. Castellan, D. B. Pisoni, & G. R. Potts (Eds.), *Cognitive Theory* (Vol. 2, pp. 21-41). Hillsdale, NJ: Lawrence Erlbaum.

Klahr, D., & Siegler, R. S. (1978). The representation of children's knowledge. In H. W. Reese & L. P. Lipsitt (Eds.), *Advances in child development and behavior* (Vol. 12, pp. 62-116). New York: Academic Press.

Kotovsky, K., Hayes, J. R., & Simon, H. (1985). Why are some problems hard? Evidence from Tower of Hanoi. *Cognitive Psychology*, *17*, 248-294.

Larkin, J. H., & Simon, H. A. (1987). Why a diagram is (sometimes) worth 10,000 words. *Cognitive Science*, *11*, 65-100.

Sleeman, D. (1982). Assessing aspects of competence in basic algebra. In D. Sleeman & J. S. Brown (Eds.), *Intelligent tutoring systems* (pp. 185-200). New York: Academic Press.

Sowder, L. (1980). Concept and principle learning. In R. S. Shumway (Ed.), *Research in mathematics education* (pp. 244-285). Reston, VA: National Council of Teachers of Mathematics.

Words into Type. (1974). Englewood Cliffs, NJ: Prentice-Hall.

ACKNOWLEDGMENT

This research was supported in part by the National Science Foundation under NSF Award Nos. MDR-84-70166 and MDR-86-00412 and by a fellowship from the John Simon Guggenheim Memorial Foundation. Any opinions, findings, conclusions, or recommendations expressed herein are those of the author and do not necessarily reflect the views of the National Science Foundation or the John Simon Guggenheim Memorial Foundation.

Artificial Intelligence, Advanced Technology, and Learning and Teaching Algebra

Patrick W. Thompson
Department of Mathematics
Illinois State University

Artificial intelligence is mentioned frequently these days, many times in the context of an advertisement that tries to sell a new expert system or expert system shell. Given the current interest but general lack of knowledge about the field, perhaps it would be useful to define what is meant by artificial intelligence.

Originally, *artificial intelligence* (hereafter, AI) was the study of how to make machines behave intelligently. By "behave intelligently," it was meant that a machine could perform a task that normally would require a human to do because the task required reasoning of some sort (McCorduck, 1979). This performance definition was soon abandoned, largely because, as a task became doable by a machine, many people took that as indicative that it did not actually require intelligence. Now, AI is defined largely by a particular programming style: Knowledge required to perform a task is represented explicitly within a program, as data. The form such data take is some discrete structure, such as a tree or digraph, or a collection of rules, like:

- If *some condition* exists (in memory), then *some conclusion* should be drawn, or
- If *some condition exists* (in memory), then *some action* should be taken,

where *some condition* and *some conclusion* are items of data or data patterns and *some action* is a procedure or procedure name.

"Reasoning" is done by a procedure that interprets knowledge-as-data and draws inferences from it; inferences so drawn are then added to the knowledge base. The behavior of an artificially intelligent program is determined by the interaction of three components: (a) the rules of inference built into the inference procedure, (b) the knowledge with which the program starts, and (c) the task itself. Of course, the *task* is itself a knowledge structure, for it is represented in the program as initial conditions and conditions to be achieved. The art of AI is to analyze tasks in such a way that one can depict knowledge which leads to correct performance under widely varying circumstances.

There should be no more mystique about AI programming than about programming in general, although, in fact, this seems not to be the case. The only major difference between AI programming and programming in general is that AI programmers have at their disposal very powerful pro-gramming languages—languages in which structure and relationship can be represented explicitly and easily, and in which there is no hard and fast distinction between data and procedure.

After saying the above, I must confess that I faced a fundamental diffi-culty in preparing this paper. Research in AI has focused, for the most part, on building programs that can solve problems, once posed, independently of human intervention (Feigenbaum & Barr, 1981). Also, AI programs tend to be quite task-oriented. While such programs may have educational and pedagogical value, it is not apparent that the value of any program extends beyond a fairly restricted task domain.

An artificially intelligent program that can solve mathematical problems may be useful as a practical tool, but giving students the capability to obtain answers per se is not a primary goal of mathematics education. It is impor-tant to affect students' thinking in a way that develops skill, but a higher-level goal is to affect students' thinking so that it holds potential for their constructing new and more powerful ways of thinking in the future (Dewey, 1945). To achieve the higher-level goal, students must come to think in terms of such things as patterns, analogies, and metaphors. Research in AI has made significant strides in explaining patterns in thought, but it has not made much headway with the issues of analogy and metaphor.

In the remainder of this paper I exercise an author's prerogative. Rather than attempt to survey the AI literature, annotating the survey with cryptic comments and leaving it to the reader to make sense of them from the originals, I instead focus on two general aspects of AI vis-à-vis mathematics education. First, I examine past research on the development of intelligent tutoring systems and give samples of a "new breed" of software. The section on intelligent tutoring systems attempts to set a tone: Though AI has much to offer, we should view its research as applied to education with as much skepticism as we would any other educational research enterprise. In the second section, I review several current projects that illustrate the power of AI concepts and methodology for developing systems that present algebraic content in substantially new ways.

INTELLIGENT TUTORING SYSTEMS

Most research connecting AI with mathematics education has been to develop intelligent tutoring systems (ITSs), which are programs that are intended to mimic the behavior of a competent tutor (see Sleeman & Brown, 1982 for one of the few collections on ITSs). At the present moment, our reaction to ITS research might be like Sam Johnson's reaction to the dog that walked on its hind legs. He remarked that we should not be surprised that the dog walks well or poorly, but that it walks at all. At that level, we can be impressed by the progress of ITS research. At a more substantive level, were I to interview an applicant for a tutoring position who displayed the heavy-handedness and lack of perspective exhibited by many ITSs, he or she would definitely not get the job. This may be more a comment on system designers' concepts of learning and teaching mathematics than a comment on the potential of intelligent tutoring systems in mathematics education.

My strongest criticism of ITS research is that, by design, there is no teacher in the picture, except perhaps as someone who sets system parameters. This is acceptable as a research ploy when one pushes a concept to see how far it can go on its own. But, as a paradigm, I believe leaving the teacher out is a mistake. It is my experience that when one designs software with the assumption it will be used by a teacher as well as by students, one designs the software so that it *supports* the teacher in improving instruction, which results in richer learning by students.

For further information on ITS research, I refer the reader to the References section (most notably, Kearsley, 1987; Sleeman & Brown, 1982; Wenger, 1987).

The Hazards of Myopia

When searching the AI literature as it relates to education, one quickly realizes the paradigmatic value given to Brown and Burton's BUGGY model of errors in place-value subtraction (Brown & Burton, 1978; Brown & VanLehn, 1981; Burton, 1982; Van Lehn, 1983). In that model, students' errors are depicted as *bugs*—"discrete modifications to the correct skills which effectively duplicate the student's behavior" (Burton, 1982, p. 157). A similar approach to analyzing errors in algebra has been taken by Matz (1980, 1982), Lewis (1981), and Sleeman (1982, 1984, 1985). In these studies, competence in algebra is characterized as the possession of a set of correct algebraic rules. For example, one rule might be:

If the goal is DISTRIBUTE,
and the current expression is of the form $A(X - B)$,
then write $AX - AB$.

Errors are characterized as manifestations of incorrect rules ("mal-rules", to use Sleeman's term).

A rule orientation is entirely natural, given the constraints of this research: that representations of knowledge be data within a program and that representations of knowledge be prescriptions for action. A rule orientation is also natural, given that the goal is to develop systems capable of solving problems posed to students, solving unanticipated problems arising from interactions with students, and determining the cause (i.e., "violated rule") of students' errors.

Aside from very serious epistemological problems with BUGGY, more pragmatic problems exist as well. BUGGY-like models of competence, whether used in place-value subtraction or high school algebra, are quite fragile and contribute less to our understanding than at first might be apparent. I say this for four reasons: (a) The principal construct of BUGGY-like models, a "rule," has not been clarified; (b) it is not clear that BUGGY-like models, in fact, model anything of interest; (c) tacit assumptions about the aims of instruction may not be valid; and (d) BUGGY-like models commonly ignore relationships between areas of mathematics.

What are rules? In AI, a *rule* is a condition-action pair of the forms shown earlier (Negotia, 1985). This use of "rule" is so general that it cannot have any psychological meaning, except as a descriptor of behavior. Rules can be used to describe reasoning (i.e., "mental behavior"), but they could just as well be used to describe the behavior of a spring under various distortions. In the final analysis, rules (in AI) are programmers' abstractions of initial and final states under some class of transformations. They tell us more about system designers than about students' mathematics.

Are BUGGY models of interest? What BUGGY-like systems model is, in effect, students' errors arising from their use of means-ends analysis to accomplish tasks of which they have no understanding and thus must imitate some observed behavior in order to produce an answer (see Erlwanger, 1973). Put another way, BUGGY-like models are models of common errors made by students who reason without meaning. BUGGY-like models may accurately describe the current state of mathematics learning, but we already know that students are prone to pushing symbols without engaging their brains. In what way does a detailed understanding of how students perform tasks mindlessly help us improve mathematics education?

Perhaps the most damaging consequence of defining competence merely as possession of correct rules is that we fail to look at incompetence as stemming from impoverished conceptualizations of material from which correct rules should have been abstracted. A student who consistently transforms $a(x + b)$ into $ax + b$, for example, probably would benefit more from instruction on order of operations and structure of expressions in arithmetic than from direct instruction on how to "distribute" across parentheses in algebra.

Assumptions about instruction. BUGGY models are fragile because they assume, overtly or covertly, that instruction emphasizes mainly the learning of procedures for making marks on paper. In place-value subtraction, for instance, when instruction emphasizes the intricacies of counting by hundreds, tens, and ones from arbitrary starting points and the relationships between counting and structured materials (e.g., Dienes blocks), one rarely sees evidence of the bugs reported by Brown and Burton (Thompson, 1982). One still sees errors, but they are more apparently conceptual—with little resemblance to the errors predicted by BUGGY.

Similarly, there are reasons to believe that, if algebra instruction emphasizes the concept of an expression as an entity having an internal structure, and if "rules" are proposed as structure-modifying transformations that leave some aspect of an expression invariant, then common errors as reported by Lewis, Matz, and Sleeman seem not to be so common (Lesh, 1987; Thompson & Thompson, 1987). I will discuss this claim again in a later section.

Relationships to other concept fields. Studies in algebra that incorporate BUGGY-like models of competence typically have not considered relationships between algebra and arithmetic. After first reading Matz (1982), I gave one of her examples to a ninth grader:

$$\text{Solve for } x: \frac{x + 1}{x + 4} = \frac{5}{6} \text{ (Matz, 1982, p. 50).}$$

The student gave the answer predicted by Matz, namely $x + 1 = 5$ and $x + 4 = 6$, or $x = 4$ and $x = 2$. I then said, as if offering a new problem, "I am thinking of a number: If you add 1 to it and then add 4 to it and make a fraction with these two numbers, the fraction reduces to five sixths," while writing the same equation as before. After listing multiples of 5 and multiples of 6, the student got the correct answer: $x = 14$. This ninth grader knew a lot about ratio and proportion, but he did not relate his knowledge to the problem when it was first presented. I grant that he did not use the procedure typically expected for solving this problem, but he did solve it, and probably with more understanding than the algorithm for "cross multiplying" requires.

I do not offer this example to "disprove" BUGGY-like models of algebraic competence. Rather, I only wish to reinforce the point that, when algebra is taught as a system for representing relationships among quantities, one sees very different, and perhaps more desirable, types of behavior than are expected under the more stereotypical circumstances assumed by users of BUGGY-like models. Moreover, if one is sensitive to the role of prior arithmetical learning in algebra and is sensitive to conceptual contexts for the formation of algebraic rules, then in designing an intelligent tutoring system for algebra, one will design the system so as to make it possible for students to reveal impoverished conceptualizations as well as buggy rules. I discuss the last point more fully in Thompson (1987a).

NEW WAYS TO SUPPORT HUMAN UNDERSTANDING

A new approach to software design is emerging from AI. The spirit of the approach is captured well in Draper and Norman's introduction to *User Centered System Design*:

> We do not wish to ask how to improve upon an interface to a program whose function and even implementation has already been decided. We wish to attempt User Centered System Design, to ask what the goals and needs of the users are, what tools they need, what kind of tasks they wish to perform, and what methods they would prefer to use. We would like to start with the users, and to work from there. (Draper & Norman, 1986, p. 2)

Norman and Draper's (1986) collection constitutes an attempt to combine various perspectives from the cognitive sciences, on topics that range from being a better typist to developing mental models of advanced technological devices, and to focus these perspectives on how best to shape the interaction between humans and computers. They note that in the pluralism of approaches there exists the germ of a new kind of interaction:

> One approach, a rather traditional one, is to start with considerations of the person, the study of the human information processing structures, and from this to develop the appropriate dimensions of the user interface. . . . Another approach is to examine the subjective experience of the user and how it might be enhanced. When we read or watch a play or movie, we do not think of ourselves as interpreting light images. We become a part of the action: We imagine ourselves in the scenes being depicted. We have a "first person experience." . . . Well, why not with computers? The ideal associated with this approach is the feeling of "direct engagement," the feeling that the computer is invisible, not even there; but rather, *what is present is the world we are exploring, be that world music, art, words, business, mathematics, literature, or whatever your imagination and task provide you* [italics added]. (Draper & Norman, 1986, p. 3)

The idea proposed by Draper and Norman—that we examine the user's subjective experience, together with emerging refinements in the technology of object-oriented programming (to be defined in the next section), to synthesize powerful, supportive computer environments—can make real what many competent teachers of algebra wish they could accomplish: *"If only I could get them to see in their minds what I see in mine!"*

I will present aspects of four projects, all currently ongoing, that illustrate the idea of "direct engagement" between students and computerized, algebraic environments. I will describe them in some detail, as the nature of the software is probably unfamiliar to a large segment of the mathematics education community. The content areas illustrated are problem solving, equation solving, concept of expression, and concept of equation.

Problem Solving

A question of critical importance in the teaching of school mathematics is how to prepare students for the transition from arithmetic to algebra. This question did not originate in AI; it has been addressed from various perspectives throughout the century (Kieran, this volume; Stanic & Kilpatrick, 1988). One of the most promising approaches is to emphasize issues of problem representation over problem solution in arithmetic problem solving (Herscovics & Kieran, 1980; Kieran, this volume). The focus of the project illustrated here is to have students represent arithmetic and algebra word problems by representing the quantities involved and relationships among them. The computer program used in this project—called *Word*

Problem Assistant (WPA) (Thompson, 1987b)—which runs on a Macintosh Plus, was inspired by a program created for a Xerox Dandelion by Valerie Shalin, Nancy Bee, and Ted Rees (described in Greeno, 1985, and in Greeno et al., 1985).

The approach embodied in WPA draws on a number of sources: AI (object-oriented systems; see Kay, 1984), cognitive science (semantic networks; see Resnick & Ford, 1981), and mathematics education. An *object-oriented* computer program is one that presents a user with what he or she identifies as objects (e.g., a graph or an image) and which possesses an internal representation of those objects and their properties. The program allows the user to act on objects in well-defined and natural ways and responds to those actions as one would predict on the basis of (perhaps informed) intuition.

A *semantic network* is a graph that depicts items of knowledge and relationships among them (Sowa, 1984). Semantic networks have been used in AI to imbue programs with "knowledge" about some domain. They have been used as theoretical constructs in cognitive science to formulate hypotheses about knowledge structures that are expressed in behavior. The novelty of the approach taken by Shalin et al., and extended here, is that the computer is used as a medium for students to externalize their knowledge structures and thereby gain the potential to refine them.

Word Problem Assistant presents students with a menu of icons with which to represent the quantities involved in a problem. There are four icons to represent four distinct kinds of quantities: numbers of things, rates, differences, and ratios. At the time of selecting an icon, the student types a natural-language description of the quantity and the unit in which the quantity is measured (see Figure 1).

The handles on the icons are a reminder of the kind of quantity being represented. The labels within cells of each icon (other than the unit cells) tell the student the information that is appropriate for entry—either a description or a value. To enter information, a student puts the pointer over the cell of interest, clicks the mouse button, and then types the information. Figure 2 shows a student's representation of the quantities in the following problem:

> A biologist released 200 marked fish into a lake. He later captured 6 samples of fish. On the average, the number of marked fish in each sample was 1/60 of the sample size. Approximately how many fish are in the lake?

To show relationships among the quantities in a problem, a student simply uses the mouse to draw arrows between them. Figure 3 shows that the quantity *No. Marked Fish* is related to *Total No. Fish* and *Mark Rate*. Figure 3 also shows a feature of WPA: Whenever there is sufficient information to fill a cell, WPA does it and puts a bullet (●) in the line to show that the

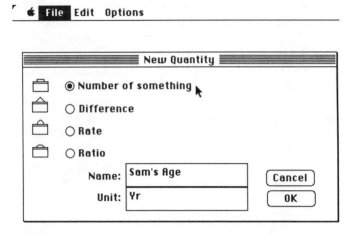

Figure 1. WPA allows natural-language description of quantities and units.

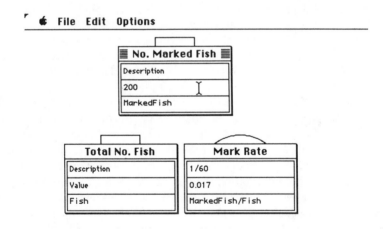

Figure 2. WPA cells after they have been filled by the user.

contents of that cell were inferred. After the student draws the arrows, WPA infers that *Total No. Fish* should be the quotient of *No. Marked Fish* divided by *Mark Rate*, since *No. Marked Fish* is the product of *Total No. Fish* and *Mark Rate*.

Figure 3 shows another feature of WPA, namely that operations are not stated explicitly. Instead, operations are implied by the kinds of units that compose the relation. The intent is that students' attention be focused on the quantities involved and relationships among them instead of on what formulas to apply. In that way, formulas are experienced as arising from an

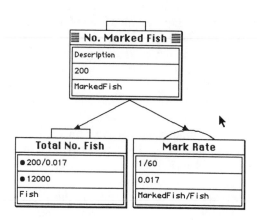

Figure 3. WPA uses a bullet (●) to show that the contents of a cell were inferred by the program.

understanding of the problem instead of as something that should be sought early on in solving a problem. This approach also is consistent with that of Kintsch and Greeno (1985), who propose that a necessary feature of a good mental representation is that it include information sufficient to determine the operations necessary to solve the problem.

To make explicit the independence of problem structure and the specific problem, WPA includes two features. First, if a value or description is changed, then all affected inferences are likewise changed. Thus, one can vary values to correspond with variations of the original problem. Second, one can change the pattern of information. For example, instead of entering values for *Mark Rate* and *No. Marked Fish*, one could enter values for *Total No. Fish* and *Mark Rate*. In this way, teachers can make explicit the idea of different problems being variations on a single theme.

WPA is also designed to make explicit the notion of a problem parameter and the distinction between a parameter and a variable. Two examples will illustrate this distinction.

Imagine this scenario: A teacher proposes the following problem to a class while using WPA with a projector or large screen monitor:

> While driving, John accelerated at 7 ft/sec/sec for 10 seconds. Afterward, he drove at a constant speed. He drove 1000 ft in the first 5 seconds after he stopped accelerating. What was John's speed as he began to accelerate?

Then the teacher guides the discussion, leading to the representation given in Figure 4.

In setting up this problem, the teacher creates icons for each quantity of the problem, plus two others not mentioned (*Added Speed* and *Final Speed*). She enters values into *Distance, Const. Speed Time, Acceleration*, and *Acceleration Time*. Then, she draws arrows as shown. WPA infers descriptions and values for *Final Speed, Added Speed*, and *Initial Speed*. After solving this problem and discussing the solution, she says: "We will have to solve another problem just like this one tomorrow, except with different numbers. But Mrs. Snodgrass will have the computer; we won't be able to use WPA. Let's generate a formula so that we can solve tomorrow's problem without having to first set it up." Then the teacher types letters into the description cells of the quantities initially known. As she types letters, WPA propagates the letters in place of the previously stored values, resulting in the representation shown in Figure 5. The letters in this representation are problem parameters, to be given values at some later time. They are not essential to the solution of any specific problem having this structure.

The distinction between variables and parameters arises from distinctions among problems in terms of the necessity of using letters to represent them. The following problem *requires* the use of letters in its representation:

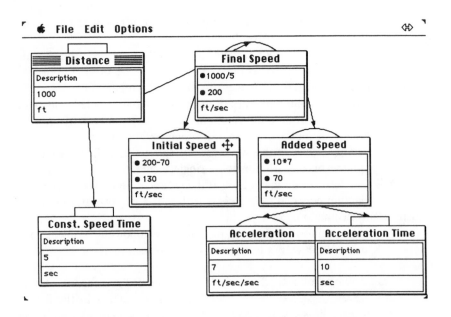

Figure 4. Final WPA representation of an acceleration problem.

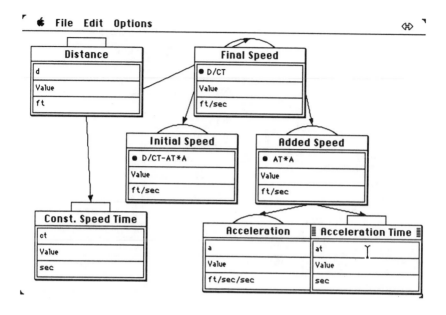

Figure 5. Use of problem parameters in WPA.

John is now four times as old as Sally. In five years, John will be three times as old as Sally. How old are John and Sally?

Figure 6 shows a representation in WPA of this problem. All values were entered; descriptions were inferred by WPA. The *Comparison 2* quantity holds the seed of the solution, as it has both a description and a value. The student can solve the problem by selecting the *Comparison 2* icon and then selecting *Solve Equation* in the Options menu.

The reason that letters are required in this problem is that, while we may assign a range of values to John's age, whatever value we assign must satisfy *two* constraints to make the network consistent. This suggests that solving problems that have the structure of simultaneous equations would be a cognitively supportive context in which to teach the concept of letter-as-variable (which would include graphing approaches to solving equations; see Kaput, this volume; Lesh, 1987).

There are many features of WPA that I would like to discuss—especially those that concern its use in instruction—but cannot for lack of space. To summarize, in traditional settings we try to communicate this message to students: A good representation of a problem is the most important step toward a solution. With WPA (and apologies to Marshall McLuhan), the medium is the message.

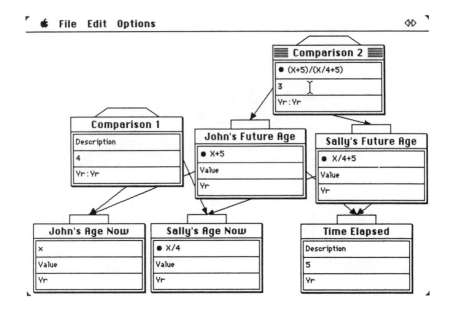

Figure 6. Use of variables in WPA.

Equation Solving

McArthur (1985) and McArthur, Stasz, and Hotta (1987) describe the first version of a system that will eventually grow into an intelligent tutor for solving algebraic equations. In its current form it does not tutor. Rather, it includes what McArthur calls "an inspectable expert." That is to say, a student can ask the "expert" to show the details of its reasoning.

Figure 7 shows a sample screen. In McArthur's system, a portion of the display shows a *reasoning tree*. A reasoning tree depicts the steps taken in solving an equation; branches from any step indicate that the student has attempted more than one solution path. The number of options available to a student is quite large, as can be seen from Figure 7. McArthur, Stasz, and Hotta (1987) give a full description of how these options are used. I will discuss only those options that make the expert "inspectable" and that support students' understanding of the reasoning in equation solving.

A student begins a problem by clicking on *New Question* (lower left side of the screen in Figure 7). After entering the initial equation to solve, the student clicks *New Line* and types a new line. Under this mode of operation, the display shows what would be shown were the student recording a solution with paper and pencil. The difference between using McArthur's system and using paper and pencil is in the editing features and checking features provided by the software. Figure 8 shows the display after the student selects *Go Back* and the equation $(-6)x = -17 + 15 + (-9)x$ as the step to go

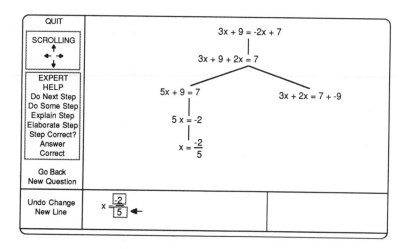

Figure 7. An adaptation of the McArthur et al. (1987) system illustrating a student's solution path.

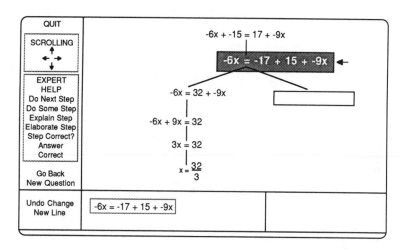

Figure 8. The system developed by McArthur et al. permits the user to return to a specific line and follow a different solution path.

back to. The program creates a branch from that step to indicate that any new equations constitute a different solution path than the one the student was originally on.

To activate the expert, the student can select a step (by clicking on a line that connects two equations) and then select *Step Correct?* from the menu options. The expert will respond "yes" or "no" to correctness, and if the step is correct, it will comment on the appropriateness of the step for achieving the goal of solving the equation. The student can also ask the expert to supply a next step, by clicking on the *Do Next Step* menu option. If the student does not understand what the expert has done, he or she can request further detail by clicking the *Elaborate Step* menu option and then clicking on the line connecting the two equations in question.

Figure 9 shows the expert's elaboration of a step. The student typed the equation $-6x = -2 + -9x$ (continuing from Figure 8). Then, he asked the expert to supply the answer. Simply $x = -2/3$ was displayed. He then asked the expert to elaborate on its solution. By clicking *Elaborate Step* and the line between the equations $-6x = -2 + -9x$ and $x = -2/3$, he was given the detailed steps for deriving the second equation from the first.

As with Draper and Norman's exhortation, we must ask what are the intended subjective experiences to be had by students who use McArthur's system. Clearly, one is that students will develop a sense of heuristic search in applying expression-transforming operations: If one path is not productive, go back to solid footing and try another approach.

A second field of experiences, based on the use of *Step Correct?*, *Do Next Step*, and *Elaborate Step*, is intended to develop the idea that knowledge itself is something that can be reasoned about.

These activities teach the student that an important part of learning a cognitive skill is *learning to study your own reasoning processes* [italics in original]. . . . When students are asked why they do poorly on a mathematics test in grade school, common attributions are "I'm dumb in math" or "I had a bad test" or "The test was unfair." Very few identify specific knowledge that they might have lacked, or even understood that their correctable

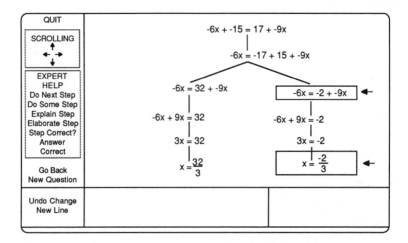

Figure 9. Right hand branch illustrates how the *EXPERT* option in the McArthur et al. system can not only provide the solution but also show the intervening steps.

lack of knowledge might have been responsible for failure. (McArthur et al., 1987, pp. 317-318)

The kinds of interactions that occur between students and McArthur's system are possible without a computer, but just barely. Even so, the modes of presentation and interaction used in this system demand the flexibility and rapidity provided only with a computer display.

Concept of Expression

In Thompson (1987a) I outlined design considerations for developing what I call mathematical microworlds—computer programs that embody a mathematical system in a way that promotes mathematical explorations. I also gave an example of a program under development that embodies an intelligent mentor to aid students in their explorations of isometric transformations of the plane. The key to incorporating an intelligent mentor into a mathematical microworld is to study students' conceptions of the subject matter *as they interact with the microworld*. By studying conceptions in the context of students' work with a microworld, one finds not only their fundamental misconceptions, but also how those misconceptions are expressed in students' interactions with the microworld. One can then design a "bug catalog" that is sensitive to qualitative, fuzzy misunderstandings. The first step toward this goal is to study students' conceptualizations.

This methodology, that is, developing the "knowledge" to be embedded in a computerized mentor by first observing students' interactions with a subject as contained in a microworld, is now being applied to the teaching of expressions and equations. The program, called *EXPRESSIONS*, is intended to emphasize cognitive and structural features of algebra that are typically not treated in standard algebra curricula, and which appear to be a source of many students' difficulties in algebra. These are:

(a) Expressions and equations have internal structure, and the structure of an expression or equation constrains what may be done to it (Greeno, 1982 [cited in Kieran, this volume]);

(b) Expressions and equations (henceforth, "expressions") are objects and, as such, they may be acted upon;

(c) A field property or an identity is a statement of relationship among an expression, the application of the property or identity, and the result of that application.

A premise that unifies the three points listed above is: Multiple representations of an object, with actions performed on one being reflected by changes in the other, facilitate students' development of relational understanding of the content (see Skemp, 1978).

In the *EXPRESSIONS* program, expressions are presented in two forms: in conventional (sentential) notation and in the form of an expression tree.

An expression in sentential notation can be displayed with or without super-
fluous parentheses, and multiplication can be represented explicitly or
implicitly.

An action upon an expression is selected by clicking the appropriate
button along the right hand edge of the computer screen (see Figure 10).
The part of the expression to be acted upon is selected by clicking on the
place within the tree that defines the to-be-acted-upon expression. Figure
10 shows the screen after a student enters $(a + b) + (c + d)$ and clicks
ASSOCL, which stands for "re-associate from left to right." Figure 11 shows

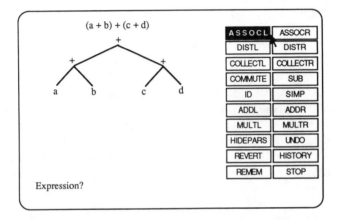

Figure 10. An *EXPRESSIONS* screen after the user entered $(a + b) + (c + d)$ and
clicked *ASSOCL*.

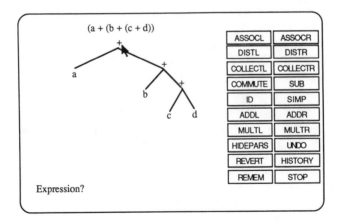

Figure 11. The resulting *EXPRESSIONS* screen after the user clicked the addition sign at
the top of the tree in Figure 10.

the screen after the student clicks the addition sign at the top of the tree, which indicates that the entire expression is to be acted upon.

To apply an identity, the student clicks *ID*, selects the identity, and then clicks the top operation of the expression or subexpression to which the identity is to be applied. Figure 12 shows the result of applying the identity $a - b = a + -b$ to the expression $b - c$.

One aim of having students operate upon expressions is to build up a repertoire of identities for future use. Any sequence of operations upon an expression produces an identity since the available operations are equivalence-preserving. When operating upon an expression, one has, at any time, a relationship of equivalence between the initial and current expressions. By clicking the *REMEM* button, the initial and current expressions are added to the list of identities and can be used thereafter. To see the deri-

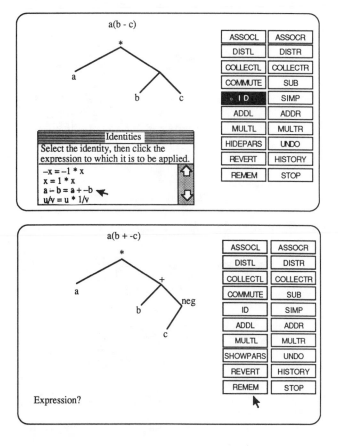

Figure 12. Two *EXPRESSIONS* screens illustrating the application of the identity $a - b = a + -b$ to the expression $b - c$.

vation of an identity, the student clicks twice on it when presented in the *Identities* window (see Figure 12).

Students can operate upon equations in any of three ways: (a) They may operate upon either side with field properties or identities; (b) they may use the buttons *ADDL* and *ADDR* to add the same quantity or expression to both sides or use *MULTL* and *MULTR* to multiply both sides by the same quantity or expression; and (c) they may replace the buttons *ADDL*, *ADDR*, *MULTL*, and *MULTR* with the single button *OPERATE*. Figure 13 illustrates the use of the *MULTL* button.

The *OPERATE* button (not shown) allows students to apply the principle that if $f(x)$ is a function and $u = v$, then $f(u) = f(v)$. Thus, if they want to square both sides of an equation and then add 3 to the result, they would

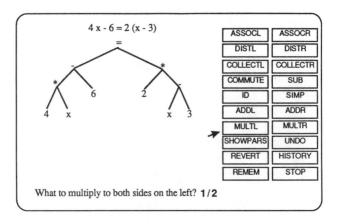

File   Windows   Options

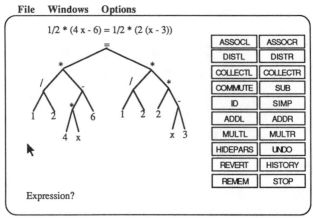

Figure 13. The use of the *MULTL* button in *EXPRESSIONS* to multiply both sides of an equation by the same quantity.

click the *OPERATE* button and type "? ˆ 2 + 3"—meaning that each side of the equation is to be composed with the function $f(x) = x^2 + 3$.

EXPRESSIONS has been used with eight average seventh graders and with a class of preservice elementary school teachers. Students' use of *EXPRESSIONS* was guided by a workbook that contained graduated activities and exercises (see Thompson, 1985, 1987a, for an explanation of graduated exercises in a Piagetian tradition). The exercises began with order of operations in arithmetic expressions, moved to derivations of equivalence of arithmetic expressions by the use of field properties, and then went on to derivations of equivalence of algebraic expressions by the use of field properties and identities.

Analyses of videotapes and computer-recorded interactions taken during the seventh-grade field test is given in Thompson and Thompson (1987). Here I will note two observations. First, contrary to my expectations, students found expression trees to be quite intuitive. In fact, they soon preferred working with expression trees to working with expressions in sentential form (all exercises were presented in sentential format).

Second, when planning the pilot study, I did not have a clear sense of how to motivate the use of letters in expressions. There was not time for an elaborated treatment, so I decided simply to begin using letters and listen closely to the students' reactions—and plan accordingly in succeeding studies. To my amazement, students were not bothered by the introduction of letters in expressions. They quickly saw letters as placeholders for substructures in a tree. For example, six of the eight students explained that in applying the associative property, $(a + b) + c = a + (b + c)$, to $(3 + 5) + (7 + 9)$, the subexpression $(7 + 9)$ "is just like c."

The fact that students saw two expressions like $(a + b) + c$ and $(3 + 5) + (7 + 9)$ as appropriate for the application of the *ASSOCL* transformation suggests that they were developing a generalized concept of variable—letters could stand for numbers but, in general, they were placeholders within a structure and *anything* could be put in their place, including other expressions. In another instance, two students, when considering what identities to apply to $(a - b)/c$ in order to obtain $(a/c) - (b/c)$, remarked that "$(a - b)$ is like u" when comparing their expression to the identity $(u/v) = (u * 1/v)$.

Frequently, students are bothered by the introduction of letters in expressions. Why? In their experience an expression is there to be *evaluated* (Kieran, this volume), and they cannot evaluate an expression having a letter in it. The seventh graders who used *EXPRESSIONS* had a different kind of experience. In their use of this software, an expression was there to be *manipulated*. The presence of a letter in an expression did not affect their ability to manipulate it.

My hypothesis about why students used substitution spontaneously when comparing a complex expression with one that had the same structure but less complexity is this: These students realized the general purpose of letters because of the nature of their activities with arithmetic expressions. When transforming an arithmetic expression, such as $(5 + 9)(5 + 9)$, to obtain some goal expression, such as $(5 \times 5) + 90 + (9 \times 9)$, the particular numbers did not matter; only the structure of the current and goal expressions constrained their actions. Also, the arithmetic exercises included many instances of transformations that left subexpressions intact, for example, transforming $3(6/9) + (6/9)7$ to $10(6/9)$. In working these exercises, students became used to ignoring, for the moment, the details of a subexpression and treating it as a single item in the super-expression. They were *thinking* in terms of substituting an expression for a variable, that is, substituting $6/9$ for x in $3(x) + (x)7$. A cognitive foundation had been laid for explicit substitution of expressions for letters.

Concept of Equation

Feurzeig (1986) has described a multifaceted project, to be used with sixth graders, that focuses on three aspects of algebraic understanding: algebra as a language for describing actions on and relationships among quantities, algebra as one instance of a class of functional languages, and algebra as the application of manipulative skills. The portion dealing with manipulative skills has resulted in a program similar in spirit to McArthur's program. In this paper, I will discuss only the aspect of Feurzeig's project that focuses on algebra as a language for describing actions on and relationships among quantities. This focus manifests itself in what Feurzeig calls the *Marble Bag Microworld*:

> The objects in this microworld are pictures of marbles and bags (bags contain some unknown number of marbles); the operations include addition or subtraction of specified numbers of bags or marbles, and multiplication or division of the current collection by a specified integer. Students are introduced to marble bag stories and diagrams. They are shown how to create and solve simple marble bag stories (story problems). They are introduced to standard algebraic notation as a rapid way of writing marble bag stories. As a student works on a problem, the system can show the correspondence between the iconic, English, and standard algebraic representations of the operations and results. (Feurzeig, 1986, p. 231)

The marble bag microworld described by Feurzeig consists of a display as shown in Figure 14. (All figures in this section were generated as screen dumps from a prototype copy of the *Marble Bag Microworld*, which was graciously lent to me by Wally Feurzeig.) The student is asked to "construct a story that ends up with the expression $2(4X) + 2$." To start, the microworld provides one bag in the *STORY* window and the line, "Think of a number . . . X," in the *History* window. Clicking three times on the bag icon in the upper left corner produces three more bags in the *STORY* window and the description of the net effect of that move in the *History* window ("Multiply by 4 . . . $4X$"). Clicking four more times on the bag icon produces four more bags in the *STORY* window ("Double what you have . . . 2(4*X*)"), and clicking twice on the marble icon produces two marbles ("Add 2 . . . 2(4*X*) + 2").

The reverse process of that shown in Figure 14 is one in which the computer generates a marble bag story in English and algebraic notation, and the student translates it into a display of marbles and bags (see Figure 15). When the story has been told, the microworld says something like, "There are now 38 marbles in all; how many marbles did I start with?" To determine the unknown number (*X*) of marbles in each bag, the student clicks on the display of marbles and bags in the *STORY* window, thereby unraveling the equation. As before, the microworld mimics the student's actions in English and algebraic notation in the *History* window, this time producing a history of the story's inverse creation. Feurzeig's intention is that these activities provide a cognitive foundation for operations on equations as ways of unraveling an equation back to its initially unknown value.

The microworld also has a "balance beam" mode. In this mode the student can operate on one side of an equation or on both sides. The balance tilts toward one side or the other if a student does something that does not maintain equivalence between the two sides. The balance beam and the unraveling approach are offered as two complementary methods of equation solving.

The most impressive feature of Feurzeig's project (not just the marble

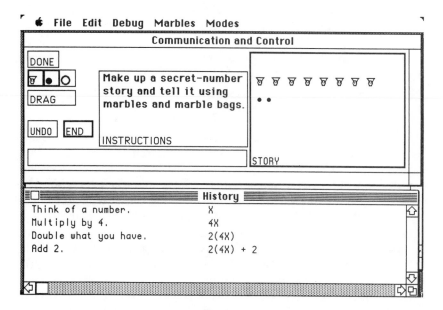

Figure 14. The student creates a secret-number story that is translated into English and algebraic notation by the *Marble Bag Microworld*.

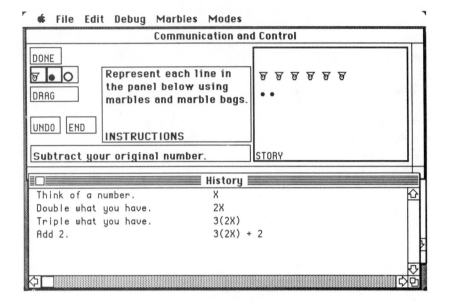

Figure 15. The student uses marbles and marble bags to represent each line in the secret-number story supplied by the *Marble Bag Microworld*.

bag microworld) is the variety of metaphors it uses to approach the idea of algebra as a language descriptive of operations upon quantities and relationships among quantities. While Feurzeig's team has yet to make the AI side of their programs apparent to an observer, the intelligence of their design decisions is readily apparent.

CONCLUSION

I began the previous section with a discussion of Draper and Norman's idea that software can provide a user with the sense of having a "first person" experience. The spirit of the pieces in Norman and Draper's collection is that of providing people with tools. The four examples of software described in this paper differ somewhat from that spirit. While one might think of them as tools for the intellect (Papert, 1980), they were designed with more in mind than aiding cognition. Rather, they were designed with the intent of *shaping* cognition. If there is an "aid" aspect to these projects, it is that the software was designed to aid students in coming to think the way the designers intended them to think.

Had I focused on tools per se, I would have had to mention hand-held, symbol-manipulating calculators recently introduced by Hewlett-Packard and soon to be released by AT&T. Had I discussed these, I would have argued that, just as with arithmetic calculators in elementary school, these machines are largely irrelevant to mathematics education unless students have appropriate mental models and knowledge structures that enable them to make judicious use of the technology.

The software examples discussed here differ significantly from previous work with intelligent tutoring systems. Developers of ITSs have tended to focus on strategic and manipulative skills within quite rigid boundaries of traditional algebra instruction. The focus of the examples presented here is at a much higher conceptual level. Were the developers of these software programs to embed their systems within ITSs, however, they could take advantage of past ITS research, especially the "issue-oriented" techniques developed by Brown, Burton, and de Kleer (1982).

The examples presented here also differ from past approaches that attempted to present mathematics as descriptive of phenomena, or as descriptive of structural features of phenomena. Rather, they attempt to present mathematics itself as a phenomenon. In a very real sense, these projects attempt to shape students' experiences so as to foster powerful, yet intuitive, conceptualizations of the subject matter. The relationship between experience and conceptualization is stronger than one might think, especially in regard to mathematics (diSessa, 1983). One could say that, in mathematics, *to experience is to conceptualize.* The critical issue, from a software design perspective, is to present an environment that allows conceptualizations at a number of levels and promotes students' advancement through those levels (Norman, 1980; Rumelhart & Norman, 1978; Thompson, 1985, 1987a; Thompson & Dreyfus, 1988).

Conceptualizations of subject matter are powerful or weak for very specific reasons. A weak conceptualization causes students to overgeneralize, undergeneralize, and do "silly" things. A powerful conceptualization provides both generality and constraints on that generality. For example, it is one thing to know a rule that transforms $a(x - b)$ into $ax - ab$, or to know that transforming $a(x - b)$ into $ax - b$ is not correct because the result should be $ax - ab$. It is quite another thing to know that transforming $a(x - b)$ into $ax - b$ is nonsense.

Experts are experts not because they do not make mistakes, but because they *notice* mistakes soon after they are made. Mistakes violate constraints. But constraints often cannot be rules, for rules are too specific. Rather, constraints are more like metaphors, such as "$(x - b)$ is a single thing, like a bucket." It is inconceivable that experts possess a myriad of "DO NOT" rules for every conceivable mistake or that they avoid mistakes by always applying rules correctly. It is much more plausible that they possess metaphors—and more broadly, conceptualizations—that allow them to judge quickly the sense or nonsense of something by the way it "fits" with their metaphoric ideas of what they are doing.

Constraints are of no use to students unless they are internalized—unless the constraints are *their* constraints. The projects illustrated here attempt to focus students' attention in ways so that they will construct metaphors, generalizations, and concepts for themselves. To what extent they succeed, and the reasons for their success or failure, are questions that need to be researched.

Another aspect that unifies the projects described here, but which has yet to become a research focus, is that of multiple, linked representational systems (Kaput, 1985, 1986). In WPA, students act on representations of problems and those actions have consequences for the expressions describing the quantities in the representation. In McArthur's system, students act on particular expressions or equations and the system fits the results of those actions into its representation of solution paths. In *Marble Bag Microworld*, students act in one representational system and the results of those actions are reflected in both the system in which the action occurred and in an alternative representational system. In *EXPRESSIONS*, students use a set of trans-representational operations to effect changes in one representational system by indicating the operand in another system—one in which the structure of the operand is made explicit and visual.

The idea of multiple, linked representational systems appears to be powerful, but we have little idea of the actual effect their use has on students' cognitions. There is preliminary evidence that their use has a positive effect on skill (Greeno et al., 1985; Lesh, 1987; Thompson & Thompson, 1987), but we do not know in any broad detail the nature of students' subjective experiences or how those experiences are reflected in cognition.

Also, we need to look seriously and vigorously at the extent to which achieving the aims embodied in these projects makes any practical difference in the mathematical lives of students. We can ask if WPA, for instance, makes students better problem solvers *away* from the program or if it, in fact, succeeds in preparing students to think algebraically. To answer these questions will not be easy. I tend to think that, before we can obtain satisfactory answers, we will need to rethink our goals and objectives for school mathematics. I suspect we will find that, yes, what students learn is valuable and should be learned, but that much of what they learn is not assessed by today's standardized tests and is not contained in today's textbooks.

The programs described in the previous section are not "big" AI programs. However, I would argue that the issues they address are actually more significant for learning and teaching algebra (or any mathematics, for that matter) than are the issues addressed by many larger projects, especially those aiming at developing intelligent tutors in algebra. This is not to say that these programs cannot benefit from past research on intelligent tutoring systems. Rather, before attempting to create intelligent tutors or intelligent mentors, we must have a better understanding of the nature of students' experiences with this type of software and a better understanding of how students' thinking can be affected when using it.

REFERENCES

Brown, J. S., & Burton, R. R. (1978). Diagnostic models for procedural bugs in basic mathematical skills. *Cognitive Science, 2*, 155-192.

Brown, J. S., Burton, R. R., & de Kleer, J. (1982). Pedagogical, natural language and knowledge engineering techniques in SOPHIE I, II, and III. In D. Sleeman & J. S. Brown (Eds.), *Intelligent tutoring systems* (pp.227-282). New York: Academic Press.

Brown, J. S., & Van Lehn, K. (1981). Repair theory: A generative theory of bugs in procedural skills. *Cognitive Science, 4*, 379-426.

Burton, R. R. (1982). Diagnosing bugs in a simple procedural skill. In D. Sleeman & J. S. Brown (Eds.), *Intelligent tutoring systems* (pp.157-183). New York: Academic Press.

Dewey, J. (1945). *Experience and education.* New York: Collier Books.

diSessa, A. (1983). Phenomenology and the evolution of intuition. In D. Gentner & A. L. Stevens (Eds.), *Mental models.* Hillsdale, NJ: Lawrence Erlbaum.

Draper, S. W., & Norman, D. A. (1986). Introduction. In D. A. Norman & S. W. Draper (Eds.), *User centered system design: New perspectives on human-computer interaction.* Hillsdale, NJ: Lawrence Erlbaum.

Erlwanger, S. (1973). Benny's conception of rules and answers in IPI mathematics. *Journal of Children's Mathematical Behavior, 1*(2), 5-27.

Feigenbaum, E., & Barr, E. (1981). *Handbook of artificial intelligence* (Vol. 1). Stanford, CA: Houristech Press.

Feurzeig, W. (1986). Algebra slaves and agents in a Logo-based mathematics curriculum. *Instructional Science, 14*, 229-254.

Greeno, J. (1985, April). *Advancing cognitive science through development of advanced instructional systems.* Paper presented at the annual meeting of the American Educational Research Association, Chicago, IL.

Greeno, J., Brown, J. S., Shalin, V., Bee, N. V., Lewis, M. W., & Yrtolo, M. (1985). *Cognitive principles of problem solving and instruction.* Abstract of Final Report to the Office of Naval Research.

Herscovics, N., & Kieran, C. (1980). Constructing meaning for the concept of equation. *Mathematics Teacher, 73*, 572-580.

Kaput, J. (1985). Representation and problem solving: Methodological issues related to modeling. In E. A. Silver (Ed.), *Teaching and learning mathematical problem solving: Multiple research perspectives* (pp.381-398). Hillsdale, NJ: Lawrence Erlbaum.

Kaput, J. (1986). Information technology and mathematics: Opening new representational windows. *Journal of Mathematical Behavior, 5*, 187-207.

Kay, A. (1984, March). Computer software. *Scientific American*, pp. 52-59.

Kearsley, G. (Ed.). (1987). *Artificial intelligence and instruction: Applications and methods.* New York: Addison-Wesley.

Kintsch, W., & Greeno, J. (1985). Understanding and solving arithmetic word problems. *Psychological Review, 92*, 109-129.

Lesh, R. (1987). The evolution of problem representations in the presence of powerful conceptual amplifiers. In C. Janvier (Ed.), *Representational systems* (pp. 197-206). Hillsdale, NJ: Lawrence Erlbaum.

Lewis, C. (1981). Skill in algebra. In J. R. Anderson (Ed.), *Cognitive skills and their acquisition* (pp. 85-110). Hillsdale, NJ: Lawrence Erlbaum.

Matz, M. (1980). Towards a computational theory of algebraic competence. *Journal of Mathematical Behavior, 3*(1), 93-166.

Matz, M. (1982). Towards a process model for high school algebra errors. In D. Sleeman & J. S. Brown (Eds.), *Intelligent tutoring systems* (pp. 25-50). New York: Academic Press.

McArthur, D. (1985). Developing computer tools to support performing and learning complex cognitive skills. In D. Berger, K. Pedzek, & W. Ganks (Eds.), *Applications of cognitive psychology*. Hillsdale, NJ: Lawrence Erlbaum.

McArthur, D., Stasz, C., & Hotta, J. Y. (1987). Learning problem-solving skills in algebra. *Journal of Educational Technology Systems, 15*, 303-324.

McCorduck, P. (1979). *Machines who think*. San Francisco, CA: Freeman.

Negotia, C. V. (1985). *Expert systems and fuzzy systems*. Menlo Park, CA: Benjamin Cummings.

Norman, D. A. (1980). Cognitive engineering and education. In D. T. Tuma & F. Reif (Eds.), *Problem solving and education*. Hillsdale, NJ: Lawrence Erlbaum.

Norman, D. A., & Draper, S. W. (Eds.). (1986). *User centered system design: New perspectives on human-computer interaction*. Hillsdale, NJ: Lawrence Erlbaum.

Papert, S. (1980). *Mindstorms: Children, computers, and powerful ideas*. New York: Basic Books.

Resnick, L. B., & Ford, W. W. (1981). *The psychology of mathematics for instruction*. Hillsdale, NJ: Lawrence Erlbaum.

Rumelhart, D. E., & Norman, D. A. (1978). Accretion, tuning and restructuring: Three modes of learning. In J. W. Cotton & R. Klatsky (Eds.), *Semantic factors in cognition*. Hillsdale, NJ: Lawrence Erlbaum.

Skemp, R. R. (1978). Relational and instrumental understanding. *Arithmetic Teacher, 26*, 9-15.

Sleeman, D. (1982). Assessing competence in basic algebra. In D. Sleeman & J. S. Brown (Eds.), *Intelligent tutoring systems* (pp. 185-199). New York: Academic Press.

Sleeman, D. H. (1984). An attempt to understand students' understanding of basic algebra. *Cognitive Science, 8*, 387-412.

Sleeman, D. H. (1985). Basic algebra revisited: A study with 14-year-olds. *International Journal of Man-Machine Studies, 22*, 127-149.

Sleeman, D., & Brown, J. S. (Eds.). (1982). *Intelligent tutoring systems*. New York: Academic Press.

Sowa, J. F. (1984). *Conceptual structures: Information processing in mind and machine*. Reading, MA: Addison-Wesley.

Stanic, G. M. A., & Kilpatrick, J. (1988). Historical perspectives on problem solving in the mathematics curriculum. In R. I. Charles & E. A. Silver (Eds.), *The teaching and assessing of mathematical problem solving*. Reston, VA: National Council of Teachers of Mathematics.

Thompson, P. (1982). *A theoretical framework for understanding young children's concepts of whole number numeration* (Doctoral dissertation, University of Georgia, 1982). *Dissertation Abstracts International, 43*, 1868A.

Thompson, P. (1985). Experience, problem solving, and learning mathematics: Considerations in developing mathematics curricula. In E. A. Silver (Ed.), *Teaching and learning mathematical problem solving: Multiple research perspectives* (pp. 189-236). Hillsdale, NJ: Lawrence Erlbaum.

Thompson, P. (1987a). Mathematical microworlds and intelligent computer-assisted instruction. In G. Kearsley (Ed.), *Artificial intelligence and education: Applications and methods*. New York: Addison-Wesley.

Thompson, P. (1987b). *Word Problem Assistant for the Macintosh* [Computer program]. Normal: Illinois State University, Department of Mathematics.

Thompson, P. W., & Dreyfus, T. (1988). Integers as transformations. *Journal for Research in Mathematics Education, 19*, 115-133.

Thompson, P., & Thompson, A. (1987). Computer presentations of structure in algebra. In J. C.. Bergeron, N. Herscovics, & C. Kieran (Eds.), *Proceedings of the Eleventh International Conference on the Psychology of Mathematics Education* (Vol. I, pp. 248-254). Montréal, Québec, Canada: Université de Montréal.

Van Lehn, K. (1983). On the representation of procedures in repair theory. In H. P. Ginsburg (Ed.), *The development of mathematical thinking*. New York: Academic Press.

Wenger, E. (1987). *Artificial intelligence and tutoring systems*. Los Altos, CA: Morgan Kaufmann.

ACKNOWLEDGMENT

This research was supported in part by the National Science Foundation under NSF Award No. MDR-87-51381 and by a grant of equipment from Apple Computer, Inc. Any opinions, findings, conclusions, or recommendations expressed herein are those of the author and do not necessarily reflect the views of the National Science Foundation or Apple Computer, Inc.

Intelligent Tutoring Systems:
First Steps and Future Directions
or
A Reaction to: "Artificial Intelligence, Advanced Technology, and Learning and Teaching Algebra"

Matthew W. Lewis
Department of Psychology
Carnegie Mellon University

My general reaction to Thompson's paper is a mixture of praise and questions. The task of reviewing and summarizing the work in artificial intelligence (AI) is unwieldy at best, and the author did a laudable job of sifting through and structuring ideas found in the literature. Thompson refers to a classic error often made in review articles: Perform a wide survey of the literature and deliver a chapter that ends up being nothing more than a cryptically annotated laundry list which makes sense only to people who know the literature. He has clearly not taken this tack and, instead, has delivered a paper focused on two general aspects of AI and algebra: (a) the relevant past work on intelligent tutoring systems (ITSs) and (b) a review of four current projects that address algebra instruction in new ways.

All in all, Thompson provides an informative review of some of the applications of AI and computer technology to algebra instruction. His definition of AI is reasonable, and the attention he focuses on the areas of intelligent behavior that AI has not yet addressed is well-directed. My discussion will address some of the main points made in Thompson's paper, with agreement and clarifications from another perspective.

CAVEATS AND DEMYSTIFICATION

Thompson's call for the demystification of programming is very important. AI is not, as some of its practitioners might like us to believe, "black magic" reserved for the understanding of a select few "enlightened" devotees. It is a set of tools, a style of programming, and a way of thinking about intelligence and approaching a task. (Winston, 1979, provides a readable introduction to AI and some of its applications; Simon, 1969, gives an in-depth introduction to some of the philosophical underpinnings of the field.) There is a tendency for people involved in AI, especially people involved in commercial applications of AI, to be lulled into a false sense of power and invulnerability, a belief that these new techniques are so powerful that the standard problems with innovation and education become unimportant. Elsewhere (Chaiklin & Lewis, 1987) I have argued that this is simply not the case. As Thompson states, we should apply the same criteria and skepticism to this new area of algebra-related research that we would apply to any other educational research enterprise or innovation.

Eliminating the Teacher

I, being an ITS researcher, take issue with Thompson's assertion that all of ITS research has been aimed at producing software that is designed to be independent of the classroom teacher. Not many ITS researchers would argue for producing "stand-alone" tutoring systems in order to take human teachers entirely out of the educational loop in K-12 settings.

In fact, what some have argued (including Lewis, Milson, & Anderson, 1987) is that the potential strength of an ITS is that it can be a powerful *tool* for the teacher if that tool fits into the "culture of the classroom." One of the promises this technology holds is that, in a class of 30 students, these systems will challenge and actively engage students in meaningful and necessary mathematical experiences—for example, exploration, extrapolation, and guided practice. However, this does not mean that the teacher sits passively and watches. Quite the contrary, while the other students are engaged with ITSs, the teacher is free to work with individuals or small groups for remediation, enrichment, or extra challenges. This interaction of teacher with tool has occurred in a six-month, public high school classroom trial of the Geometry Tutor, an ITS that teaches geometric proof (Anderson, Boyle, & Yost, 1985).

Much of ITS research has worked toward the goal of producing tutoring systems that approach the effectiveness of a quality, personal human tutor. If we think of the most gifted teacher we know, would we not like to somehow replicate that teacher and give a clone to each student for some individual tutoring during a class? This is particularly true in the case of students who are having difficulty learning, for whatever reason. There is undeniably a tension between the goal of having individual tutors for each student—in order to maximize learning—and other important educational goals, such as cooperation, socialization, role-model presentation, emotional growth, and motivation, among others. Devising an ITS that Thompson would "hire" is a goal that is a very long way off and may well be unachievable. That should not, however, stop us from using it as one of our criteria for research. As Thompson points out, the potential educational power of ITSs is still unknown.

The Myopia of Rule Orientation

Taking a rule-oriented approach to modeling students (more generally, taking an "information processing approach" to the study of cognition) has yielded interesting results, both in the psychology of how people solve problems and also in the field of AI. Unfortunately, I cannot recall the source of the aphorism that sums up my feelings about a rule-oriented approach:

A good model is one that holds together long enough to get you to a better model. My version of this is, "Push it 'til it breaks." Despite their obvious limitations, rules can do more than simulate "mindless" behavior (in our case, symbol manipulation of algebraic expressions). For example, rules are used in systems that reason about, diagnose, and then tutor the process of diagnosing an infection (Clancey, 1982). In McArthur's system (McArthur, Stasz, & Hotta, 1987), discussed in Thompson's paper, rules set strategic goals like getting rid of parentheses in order to simplify an equation. But, the system can explain this reasoning, which raises the question of defining what "mindless" and "meaningful" really mean.

Thompson asks, "What are rules?" *Rules* are formalisms that have proved useful in modeling a wide range of "intelligent" behaviors and mental processes. In the work on algebra, the use of rules has not been extended to the mathematical reasoning that underlies the behavior of solving novel problems. This does not mean it cannot be done. (Here I begin to balance the desire to avoid making claims that we, as researchers, cannot uphold with the desire to project where the research is heading.) "Understanding" is poorly understood. How the semantics of mathematical knowledge underlies and influences the behaviors of problem solvers is a relatively unexplored research area. We now understand certain classes of errors, as Thompson points out. These may be the "uninteresting" errors, but we are taking one step at a time. The interesting questions indicate how hard the next step will be.

Simulating problem-solving behavior. Thompson's description of a student's approach to solving

$$\frac{x + 1}{x + 4} = \frac{5}{6}$$

was fascinating. What *was* that student doing? Was it magic? The assumptions of the information processing perspective argue that there were cognitive processes taking place. Roughly speaking, these processes included noticing features of the problem and keeping them in mind while manipulating symbols related to those features to produce some problem-solving actions, both mental and written. It would have been interesting and enlightening to have asked the student to think out loud while he solved the problem (Ericcson & Simon, 1980). The problem-solving processes of this student might then have been simulated by a set of rules (Brownston, Farrell, Kant, & Martin, 1985). It is also important to note that there are other research tools for simulating problem-solving behavior, for example, parallel distributed processing models (McClelland & Rumelhart, 1986), that are being offered as alternatives to rule-based models.

Buggy models. Finally, Thompson's points on Buggy-like models are very well taken. Conceptual contexts underlying the formation of rules are very important and poorly understood. The question of how to represent this knowledge appropriately is critical and one that is at the forefront of research in both AI and cognitive psychology. The rule-based simulations that are the core of some current ITS research and development are impoverished and incomplete. However, they are a first approximation and, by developing them, we have learned a great deal about building theories of human learning.

Summary of Projects

Thompson summarizes several nice pieces of work on how the potential of advanced technology has been creatively applied to algebra instruction. Throughout, he emphasizes that there are many interesting questions that underlie further developments. Just what is this "algebraic understanding" we want students to acquire? How can we facilitate the processes of metaphor and generalization? What underlies the power of multiple, linked representations as instructional tools? What do students come away with after experiences with these systems? Does the understanding developed with these systems transfer to pencil and paper or other domains? Unanswered questions abound.

Overall, the theme of the paper is that there are some nice examples of people doing interesting work and there are many important questions to be answered. The key issue is how to represent conceptual knowledge and its links to problem-solving skill. In addition, Thompson's call for skepticism and the demystification of AI and AI programming is well stated and should be taken to heart by all, including AI researchers.

REFERENCES

Anderson, J. R., Boyle, C. F., & Yost, G. (1985). The geometry tutor. In *Proceedings of the International Joint Conference on Artificial Intelligence* (Vol.1, pp. 1-7). Los Altos, CA: Morgan Kaufmann.

Brownston, L., Farrell, R., Kant, E., & Martin, N. (1985). *Programming expert systems in OPS5: An introduction to rule-based programming*. Reading, MA: Addison-Wesley.

Chaiklin, S., & Lewis, M. W. (1987). Will there be teachers in the classroom of the future? In J. Jacky & D. Schuler (Eds.), *Proceedings of the Directions and Implications of Advanced Computing '87 Symposium* (pp.115-124). Seattle, WA: Computer Professionals for Social Responsibility.

Clancey, W. J. (1982). Tutoring rules for guiding a case method dialog. In D. Sleeman & J. S. Brown (Eds.), *Intelligent tutoring systems* (pp. 201-225). New York: Academic Press.

Ericcson, K. A., & Simon, H. A. (1980). Verbal reports as data. *Psychological Review, 87*(3), 215-251.

Lewis, M. W., Milson, R., & Anderson, J. R. (1987). The TEACHER'S APPRENTICE: Designing an intelligent authoring system for high school mathematics. In G. P. Kearsley (Ed.), *Artificial intelligence & instruction*. Reading, MA: Addison-Wesley.

McArthur, D., Stasz, C., & Hotta, J. (1987). Learning problem-solving skills in algebra. *Journal of Educational Technology Systems, 15*(3), 303-324.

Research Issues in the Learning and Teaching of Algebra

McClelland, J. L., & Rumelhart, D. E. (1986). *Parallel distributed processing* (Vol. 2). Cambridge, MA: MIT Press.
Simon, H. A. (1969). *The sciences of the artificial*. Cambridge, MA: MIT Press.
Winston, P. H. (1979). *Artificial intelligence*. Reading, MA: Addison-Wesley.

ACKNOWLEDGMENT

This research was supported in part by the National Science Foundation under NSF Award No. MDR-84-70337. Any opinions, findings, conclusions, or recommendations expressed herein are those of the author and do not necessarily reflect the views of the National Science Foundation.

Linking Representations in the Symbol Systems of Algebra

James J. Kaput
Department of Mathematics
Southeastern Massachusetts University
and
The Educational Technology Center
Harvard Graduate School of Education

Mathematics is, among other things, a collection of languages, and languages have dual, interlocking roles: They are instruments of communication, and they are instruments of thought. In a previous paper (Kaput, 1987b) I developed a beginning set of languages for communicating and thinking about mathematical languages, which in algebra include expressions, equations, coordinate graphs, tables of data, hybrid natural language fragments, and other notations. In this paper I will apply and extend the previously developed terminology and theoretical framework to discuss and evaluate the representational characteristics of new or potential algebra learning environments, especially those that support the *linking* of different algebraic representations.

Thus this paper is deliberately oriented towards future research, indeed towards the research needed to design the future. To consider research from any other perspective would contribute to the well-documented and continuing failure of school mathematics to serve students' real needs.

FOUNDATIONS

Our starting points are: (a) the notion of *mental representation* as the means by which an individual organizes and manages the flow of experience and (b) the notion of *representation system* (or *symbol system*) as a materially realizable, shared cultural or linguistic artifact.

Representation systems, when learned, are used by individuals to structure the creation and elaboration of their own mental representations. A mathematical representation system is used to instantiate mathematical objects, relations, and processes in material form—the physical signs produced by pen on paper, keystroke on computer, and so on. A fuller discussion of representation systems and the underlying issue of Platonism can be found elsewhere (Kaput, 1985, 1987a, 1987b, in press).

A caveat: Our initial attention to symbol systems should not be misread as an assertion that mathematics is, and hence the curriculum should be, about symbols and syntax. On the contrary, our ultimate aim is to account for the building and expressing of mathematical meaning through the use of notational forms and structures. One of the problems to be addressed by algebra research is that of student alienation, which, I suggest, is the result of teaching algebra syntax instead of semantics. This problem is compounded by (a) the inherent difficulties in dealing with the highly concise and implicit syntax of formal algebraic symbols and (b) the lack of linkages to other representations that might provide informative feedback on the appropriateness of actions taken (Kaput, 1987c).

Four Sources of Meaning in Mathematics

As meaning is the foundation of mathematics learning, it is the foundation of this paper. Although I will have more to say about the notion of mathematical representation system later, for purposes of these prefatory remarks the reader can assume such systems include the familiar systems of coordinate graphs, algebraic equations, and so on. In these terms I assert that mathematical meaning has at least four sources, grouped in two complementary categories:

(a) *Referential Extension*:
   1. Via translations between mathematical representation systems;
   2. Via translations between mathematical representations and non-mathematical systems (including physical systems, as well as natural language, pictures, etc.);

(b) *Consolidation*:
   3. Via pattern and syntax learning through transformations within and operations on the notations of a particular representation system;
   4. Via mental entity building through the reification of actions, procedures, and concepts into phenomenological objects which can then serve as the basis for new actions, procedures, and concepts at a higher level of organization (often called "reflective abstraction").

School mathematics tends to be dominated by the third form of meaning building, the narrowest and most superficial form.

The study of meaning sometimes comes under the heading of "semantics." Ours is a *relational* semantics. That is, we do *not* assume the existence of absolute meanings or absolute sources of meaning. Rather, meanings are developed within or relative to particular representations or ensembles of such. Thus, for example, there is no absolute meaning for the mathematical word *function*, but rather a whole web of meanings woven out of the many physical and mental representations of functions and correspondences among representations.

Some Definitions

It is useful to be a bit more explicit about our meanings of *meaning*. In terms of Figure 1, the "meaning of A" (say, a polynomial expression) may be provided by a graphical referent B, in which case we would say "A refers to B," that is, "$16x^2$ refers to the parabola." We can also say that A *represents* B; and when A has a syntax and there is a well-defined correspondence between A and B coordinating the syntax of A with the structure of B, then we often refer to A and B and the correspondence between them as comprising a *representation system*. When A and its syntax are considered apart from a field of reference, they are called a *notation system* (see Kaput, 1987b, in press, for further details).

But all of this occurs in two worlds, in the essentially private world of mental events and state changes (the upper part of Figure 1), and in the public world of characters and lines in some physical medium (the lower part). In fact the *direction* of the reference usually depends on the cognitions of the symbol user in particular cases. Further, it is often the case that B, in turn, refers to some C, where C, the particular referential meaning of B, may be an aspect of some non-mathematical entity or situation, such as the velocity of a falling body. In this case we say "A [and/or B] is a model of C," in the usual sense. Of course, there may be many A_{cog}s for a given A, even within a particular individual, and there may be more than one A representing a particular B.

Problems of moving from the lower to the upper part of Figure 1, of constructing personal meaning, are the traditional problems of learning and understanding. Problem solving usually involves cycles between the upper and the lower parts of the diagram. It is my hope that research in algebra learning and understanding can begin to deal in more systematic ways with the problem of meaning building. I will begin my own analysis with the traditional representation systems of algebra and then turn to elaborations of these, as well as to novel systems that dynamic computer media make

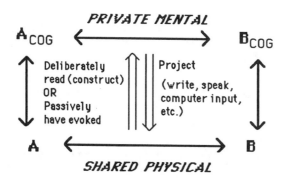

Figure 1. Correspondences between mental and physical representation systems.

possible. As the title of the paper implies, I am especially interested in how representations may be profitably linked—in both the physical and mental worlds of representations.

THE THREE CORE REPRESENTATION SYSTEMS

In keeping with the terminology developed in earlier papers, we shall let E-2 denote the set of two-variable algebraic equations in x and y with real coefficients, G-2 the corresponding set of curves in the x-y plane, and T-2 the set of two-column tables of ordered pairs of real numbers. By further specifying characters and syntax (which requires defining the well-formed objects and allowable actions on them), we obtain the three notational systems commonly used in algebra—equations, graphs, and tables of ordered pairs. When associated with a reference field, each notational system becomes a representation system. The same applies to B-2, the set of binary relations on real numbers, given in one of the standard set notations. Any of these notation systems can act as a reference field for any of the other systems or as a model of some quantitative aspect of an event, entity, or situation.

Until otherwise indicated, we will assume that each system is instantiated in a static, two-dimensional, two-colored medium—for example, paper and pencil.

B-2 and Models

There is a strong tendency in the 20th-century mathematical community to use abstract, feature-bare notations to denote "real" mathematical objects and to use feature-rich notations in applications. Thus, from a mathematical point of view, the more "abstract" B-2 is usually thought of as occupying a privileged position in that the other systems are regarded as representing *it* (see Figure 2).

The usual set-based descriptions possess virtually no qualities not also embodied in E-2, T-2, and G-2. The set-based descriptions, bare as they are, are useful for emphasizing the common role of ordered pairs in the other notations. On the other hand, the richer features of the other systems can support the building and interrelating of cognitive structures. More generally, one can think of the features of richer notations as mediating between the more abstract notations and the infinitely varied features of the world they model.

As shown in Figure 3, the variety of situations C which can be modeled by any of the four representations depends on how the features of each match with those embodying the quantitative relationships of the situations being modeled. Typically, one reads a description of a situation (a "word problem") and builds a cognitive representation of that situation in which certain features carry the quantitative relationship to be represented. These

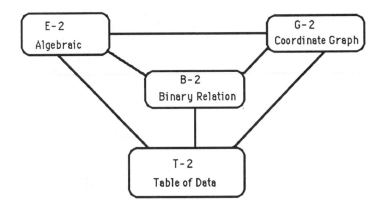

Figure 2. Interrelationships among various two-variable algebraic notational systems.

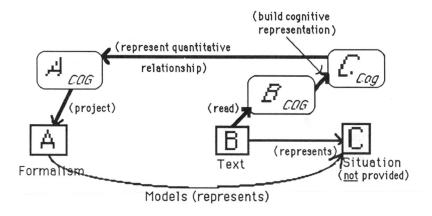

Figure 3. Building a cognitive representation system to model a physical situation.

features evoke another cognitive representation that is then projected onto an external representation (see Figure 3), a process often repeated in cyclical fashion.

The critical issue is the availability of features in the representing system to correspond with those of the situation being modeled. It is here where *B*-2 is especially weak compared to the other systems. This view is very close to Freudenthal's (1983) phenomenological view, but with more explicit emphasis on the *forms* of the representations that evoke the phenomenal experience of mathematical concepts. As is suggested by Figure 3, the process of applying the formalism is closely related to the process of learning the formalism: Relationships among the features of the formal system can be learned first in the context of situations modeled. The pedagogical problem is to render the resulting A_{cog} free enough from the starting contexts to apply it to other situations (Filloy & Rojano, 1985). This is where multiple, linked representations are especially useful (Kaput, 1987c).

Differences Among the Three Core Representations

Both tables and graphs display binary relations but differ considerably in the way they sample the domain of a relation. Tables display discrete, finite samples, whereas coordinate graphs display continuous, infinite samples—a much fuller presentation. Moreover, a coordinate graph consolidates information by condensing pairs of numbers into single points; so to compare two pairs, one need only examine the relative position of two points rather than relations among four numbers.

A critical difference between *T*-2 and *G*-2 is that a graph engages our gestalt-producing ability, which allows us to consolidate a binary (and especially a functional) quantitative relationship into a single graphical entity—a curve or line. Usually a sophisticated act of inference is required to produce a conceptual entity from a table of data. But once a conceptual entity is available, one can reason with it (Greeno, 1983). We have seen very powerful consequences of this simple fact in the context of ratios represented as slopes of lines (Kaput, 1987e; Kaput, Luke, Poholsky, & Sayer, 1986). For example, one can compare the size of ratios by comparing the slopes of their lines.

While graphical notations enable one to bring to bear much previously learned knowledge about such things as lines, curves, parallelism, intersections, and steepness, this advantage has a negative side as well. Visual knowledge and inference processes from everyday experience can be misleading in coordinate graph contexts (Goldenberg, 1987), although newer computer display systems help compensate for the shortage of contextualizing features by, for example, providing several scalings of a graph simultaneously. A table of data displays information more specifically quantitative in character than a coordinate graph, especially if the data in the table are linearly ordered and evenly spaced. Changes in the values of variables are relatively explicitly available by reading either horizontally or vertically.

Another difference between *G*-2 and *T*-2 is that the quantities involved are all *automatically* ordered in *G*-2, but not in *T*-2. The user (or some other agent, such as a teacher or computer) must provide the ordering in *T*-2. This difference illustrates the important point that some notations supplant cognitively significant or difficult actions that are the responsibility of the user in other notations (Salomon, 1979).

Such imperfect correspondences between different notations, while contributing to difficulties in translation (especially when they are not the subject of explicit instruction or attention), also serve to increase the utility of any mathematical ideas that they represent—different features apply differently in different situations. Imagine the uselessness of a concept of function represented *only* as a set of ordered pairs (Kaput, 1979)!

Differences in symbol systems also affect the *learnability* of mathematical concepts—linearity, for example. In *E*-2 the equation $y = 2x + 3$ embodies certain critical features of the quantitative relationship between x and y very explicitly and compactly. It is a shorthand rule (not the only one, of course) that generates y values from x values. Such procedural knowledge is not easily inferred from tables of data or coordinate graphs. The equation also has a feature (the coefficient of x) that conveys conceptual knowledge about the constancy of the relationship across allowable values of x and y—a constancy inferable from the table only if it is ordered and includes a full interval of integers in the x column so that one can compute "first differences" in the y column. More generally, parameters in equation notation can serve the modeling process by providing explicit conceptual entities to reason with.

On the other hand, the strictly alphanumeric representation of variables is very *implicit* regarding variation across the domain set. One reason the concept of algebraic variable has been so difficult for students to learn is that its alphanumeric representation is so implicit. The *user* must, in a real sense, supply the variation. This is less true in *T*-2. And in *G*-2, many values of the independent and dependent variables are provided simultaneously, in compact form. To provide the variation, all the user needs to do is trace a finger or in some other way follow along the graph, thereby reinvesting the spatial notation with the time dimension to provide the temporal variation that it has captured and frozen in place.

Display Notations Versus Action Notations

There is yet another deep difference among these notations based on how they are *used*. Some representations are used mainly to display, while others are designed to be acted upon.

Display notations can display procedures, relationships, or the results of actions. In a sense, all notations are display notations, but I will reserve this designation for those that do not directly support user-executable actions. Examples include tables of data and coordinate graphs. We do not traditionally act on these notations except to build them initially. Their chief function is to display information and relationships.

Some display notation systems have objects that represent procedures of one kind or another. We refer to these as *procedure-objects*. The objects of *E*-2, which are equations, represent statements about procedures for operating on numbers and comparing the results of those operations, hence are procedure-objects. Similarly, standard polynomial expressions are procedure-objects. It is possible for a display notation to include procedure-objects that are not intended for manipulation by a user. Such is the case with an x-y two-column table representing a function, where the function is explicitly described in closed form above the y column. Here the function describes a procedure for generating the right column from the numbers in the left column, but, as with the numbers in the columns, it is intended for display, not manipulation.

Some types of procedure-representing objects are more explicitly procedural than others. For example, a polynomial expression is a rather implicit representation of a procedure for acting on numbers compared to a function machine, especially a function machine that explicitly represents all the component terms as (sub)machines together with the concatenating operations.

Action notations, on the other hand, support a variety of transformations and other actions on their objects. In effect, they capture the time dimension in the user actions they support. For example, the alphanumeric form of equations in E-2 supports user transformations. Such transformations can yield equations that are semantically equivalent (have the same solution sets) or not. However, the key is that this type of notation directly supports read-write-read sequences (or "visually moderated sequences" in the sense of Davis, 1984) with rules governing semantics-preserving actions.

Actions can include syntactic transformations but can also involve other actions that modify the notation, such as substituting values for the variables of an expression or equation, building or extending a table of data, changing the parameters of a function in order to observe changes in its graph, "zooming in" on a small interval of a graph in order to inspect some local phenomenon (Tall, 1985), and so forth.

Particularly useful are representation systems with action notations that are coordinated with their reference fields, so that there is a visible correspondence not only between *objects* of the two notations involved, but between the *actions* as well. As will be illustrated below using equations and graphs, this type of representation situation is much more accessible in dynamic media, such as computers, where multiple windows on a single screen allow actions taken in one window to be immediately and saliently displayed in another.

It should be noted that the distinction between "display" and "action" notations does not refer to absolute properties of systems, but rather to types of features that evince different patterns of use. Thus, in particular cases a system that supports actions may be used only for display purposes, and vice-versa.

The medium in which an action notation system is instantiated has a great bearing on the kinds of actions that may be supported and the way that consequences of those actions are displayed. For example, if a function machine were instantiated in a paper-and-pencil medium, then the evaluation process would not be directly represented, whereas in the computer medium, one might be able literally to put numbers into the input gates (by dragging and depositing, or simply typing in) and then watch them (as bulges in a wire or mice swallowed by a snake) move along the lines connecting the machine's components until they appear transformed at the output gates (Kleiman, 1986). As described later on, other software environments alter the display/action character of some traditional notations by supporting new forms of actions, for example, direct actions on coordinate graphical displays.

Syntactic Versus Semantic Actions

Actions on, or elaborations of, a representation occur in two broad classes:

1. A *syntactic action* involves manipulating the symbols of the representation, guided only by the syntax of the symbol scheme rather than by a reference field for those symbols. (A syntactic action is an action on a notation system rather than on a representation system.) Solving the equation $x + 7 = 12$ syntactically would involve subtracting 7 from both sides, performing the associated arithmetic, and reading off the answer. A syntactic transformation treats a polynomial expression or equation as a manipulable character string subject to certain constraints.

2. A *semantic action* (or *elaboration*) is guided by the referents of the symbols, as when one solves $x + 7 = 12$ as a statement about addition of numbers—to determine, perhaps by sequential trial and error, what number added to 7 yields 12. Here the elaboration is being guided by the features of a reference field for the symbol system rather than by its syntax (using the syntax implicitly). A semantic elaboration treats an equation as a comparison of two statements about numbers (the right side happens to be degenerate here). It is important to realize that a given equation may support several different semantically guided actions, depending on the different reference fields employed and how they are linked to it.

The syntactic/semantic distinction is meant to delineate polar extremes, with the realization that most symbol-use acts involve a mixture of the two. Indeed, much real mathematical activity can be regarded as transactions between a notation system and its reference field. In the strict sense of referential meaning, our distinction is between referentially meaningless syntactic actions and referentially meaningful semantic actions. In Bruner's (1973) terms, syntactic actions are "opaque" uses of the symbols, whereas semantic actions are "transparent" uses, in which the user "sees through" the notation to be guided by the field of reference. An individual's approach to a task will likely move from one type toward the other over time, depending on intervening instruction and/or practice. There may even be a natural progression from semantic toward syntactic elaboration as the symbols and actions associated with them "reify" into entities that, in turn, can serve as referents for new symbol systems—the fourth source of meaning listed earlier (see Kaput, 1987b, 1988b, for discussion and examples illustrating the key role of symbolization in the "objectifying" process).

And, of course, one of the keys to the power of mathematics is its support of syntactically based transformations. The processing effort is off-loaded onto the machinery of the notation system when it is cognitively impossible to manipulate the referents with the necessary facility. Another key is that the beginnings and endings of these manipulations can be anchored in the field of reference. The increasing availability of technological aids to syntactic manipulation heightens the importance of this issue. But whatever the technology, *the bird of mathematical competence cannot fly on one wing*— neither the syntactic nor the semantic suffices alone. By providing the means to link actions on a notation with their consequences in a reference field, new technologies may help redress the semantic/conceptual balance without giving up syntactic/procedural power.

NEW REPRESENTATIONS AND NEW LINKAGES IN THE COMPUTER MEDIUM

Most of the mathematics related to algebra was constructed under the tacit assumption of inert, static media and expensive computation, expensive in terms of cognition and time. Indeed, one way to view the representations and the transformation/translation techniques of algebra is as a response to those constraints, neither of which any longer obtains.

Dynamic media can now support, for example:

(a) direct computational support for transformations of the traditional alphanumeric representation systems (e.g., symbol manipulators and root finders);

(b) within-representation structuring of and feedback on actions, including expert assistance (e.g., intelligent tutoring);

(c) new dynamic linkages between notations, both new and old;

(d) new actions on old notations;

(e) the generating of new dynamic notations, including notations that reveal the structure of processes in newly explicit ways;

(f) dynamically manipulable simulations of phenomena with quantitative content; and

(g) the capturing and generalizing of actions into repeatable, nameable, and inspectable procedures.

The first of these examples has dominated the nontrivial use of computers in *using* algebra, while the second has dominated the nontrivial use of computers in *learning* algebra, especially transformations within the alphanumeric notation system. (I characterize CAI "workbook" styles of computer use as trivial, so they are not considered here.) The last five types of support (c-g) are less established in current practice but are likely to dominate the applications of information technology in the not-too-distant future, especially in combination with the first two. These new modes of usage can be characterized as changing in fundamental ways the notations and actions that are used to represent mathematical relationships and processes (Kaput, 1986).

Given space constraints, I cannot give a detailed account of each new type of computer use but will, instead, give brief accounts of most of them. Each raises its own collection of research questions, and the set of such collections forms a constellation of issues needing immediate attention in the mathematics education research community.

Listed below for convenient reference are short names for the five standard notation schemes that will be discussed. Details of their syntax in different instantiations will be described later.

- *E-N*: *N*-variable equations in alphanumeric form.
- *F-1*: Functions of a single variable in the form $y = \dots$ (i.e., elements of *E-2*).
- *G-N*: *N*-dimensional coordinate graphs (for $N = 1$, this is the number line).
- *T-N*: An *N*-column table of data.
- *B-2*: A set-based, ordered-pair description of a binary relation.

The Dynamic Linking of Notations

This type of computer use may in the long run turn out to have the widest application. Whereas static media can normally support only clumsy, serial translations across notation systems, executable in real time, new computers can support, via appropriate windowing and interfacing, multiple, linked notations. (Note that notation schemes, taken in pairs, constitute representation systems, by our definition, so we also use the phrase "multiple, linked representations" to describe this genre of computer use.) The computer can be used to render the correspondence between notations immediate, explicit, and salient, where the counterparts of actions taken in one notation can be exhibited in others in a time-independent way (Kaput, 1988b).

Some of what is said below about linked representations draws on recent work in algebra and ratio reasoning at the Educational Technology Center, which has been operationalizing these assertions in particular software environments (some of which are being developed at an ETC Consortium member, the Education Development Center). However, rather than describing particular software, I will describe *styles* of software environments that may differ in various ways from those being developed. First, let us consider an example to set the stage for some more general observations about multiple, linked representations.

Action links between functions and graphs. The first environment we shall consider combines (a) a student-controlled symbol manipulator that acts on single-variable expressions (hence on functions of a single variable) and (b) a function grapher that simultaneously graphs the expressions in an adjacent window (in the form y = expression). Actions in the coordinate graph representation *G*-2 are driven by actions on the alphanumeric representation, *F*-1. For example, a legal algebraic transformation on an expression (e.g., properly expanding the product $[x + 3]^2$ to $x^2 + 6x + 9$) produces no change in the graph because the new expression represents the same function, whereas a transformation that changes the function to, say, $x^2 + 9$ causes a change in the graph, as shown in Figure 4.

A utility then enables the student to plot the *difference* between the original expression and the one that has been improperly generated (in this case $6x$). This provides rich feedback regarding the nature of the manipulative error. Another part of the software environment enables the student to experiment with alphanumeric representations of the plotted difference to determine, if possible, its polynomial representation.

Typical activities supported in this type of environment include (a) transforming a given expression into a target expression, where the given and

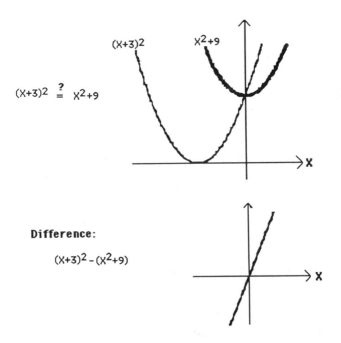

Figure 4. Screen dump showing an illegal alphanumerical transformation reflected in a coordinate graph representation, including a plot of the difference function.

target expressions are both graphed, or (b) simplifying or expanding expressions, with the feedback information being provided graphically as described above.

This environment links not only the notations *F*-1 and *G*-2 so that functions of a single variable correspond to their graphs in the usual way, but it also links *actions* on *F*-1 to their consequences in *G*-2. Transformations that preserve semantic equivalence are those for which the target and starting expressions have the same graph. Here the traditional action/display character of each representation is preserved. Graphs display, while expressions are acted upon. The critical difference, however, is the "hot action link" to the graphs. *G*-2 *can now display directly the results of actions taken in F-1.*

We need to test systematically the hypothesis that this kind of environment can provide novel and potent means for remediating many of the classic algebra manipulation errors that have been studied in recent years (e.g., by Matz, 1980).

Power of multiple, linked representations. With the above example in mind, a few general remarks are in order regarding the cognitive potential of dynamic links between representations.

What is a major source of the referential (as opposed to expressive) accuracy of natural language in the face of its inherent ambiguity, nuance, and richness? One answer lies in its redundancy. But multiple, linked representations of mathematical ideas likewise provide a form of redundancy, a redundancy that can be exploited directly in a multiple, linked representation learning environment.

Another source of power is that multiple representations enable us to suppress some aspects of complex ideas and emphasize others, thereby supporting different forms of learning and reasoning processes. With more than one representation available at any given time, we can have our cake and eat it too, in the sense of being able to trade on the accessibility and strengths of different representations without being limited by the weakness of any particular one (Kaput, 1988b).

Exactly this representational opportunism is being used in an ETC Ratio Reasoning Project (Kaput & Gordon, 1987; Kaput, Luke, Poholsky, & Sayer, 1986, 1987) where the learning environment is intended to "ramp students upward" from concrete iconic notations to abstract and flexible notations (such as *F*-1, *T*-2, and *G*-2). Actions are supported in any of the several notations, with consequences inspectable in any of the others. In this type of environment the computations required to translate actions across representations are done by the computer, leaving the student free to perform the actions and monitor their consequences across the representations.

But, perhaps most importantly, *the cognitive linking of representations creates a whole that is more than the sum of its parts*, in the same sense that binocular vision is more than the simple sum of perspectives. It enables us to "see" complex ideas in a new way and apply them more effectively. Ongoing ETC work suggests that appropriate experience in a multiple, linked representation environment may provide webs of referential meaning missing from much of school mathematics and may also generate the cognitive control structures required to traverse those webs and tap the real power of mathematics as a personal intellectual resource. However, one must choose such representations carefully and, even more importantly, ensure that the learning environment supports rich sets of actions that will expose underlying invariances and thus enable the student to weave a flexible and enduring web of mathematical meaning.

Changes in the Meaning of the Phrase "To Solve an Equation"

Since the early days of algebra as we know it (since Viète), "to solve an equation" has meant to use the available syntactic methods to transform the expressions in an equation until the resulting representation makes the roots of the equation cognitively accessible. Modern technology has helped mainly by providing syntactic assistance (via symbol manipulation software), when closed forms are desired, or numerical methods, when approximations are appropriate. Neither of these technological interventions has yet had much impact on the average algebra classroom, despite some extremely persuasive demonstrations in recent projects that these aids can free the student (and the curriculum) to focus on realistic applications of school mathematics (Fey, 1984, this volume; Heid & Kunkle, 1988). Nonetheless, the emergence of technological interventions that are "friendlier" may change the curricular status quo.

Because solving an equation amounts to elaborating a representation to render explicit the values for which the equation is conditionally true, solving an equation is very much representation-dependent. Both the elaborations and the display characteristics of the representations determine what one can do and what one can "see." So we will examine equation solving in different representational contexts.

Algebraic table-of-data interpretation. Fey, Heid, and colleagues have developed a rather direct extension of the muMath symbol manipulation package to include a table-generating feature (Heid & Kunkle, 1988). Heid has extended the table-of-data representation to include both the automatic generation of entries for more than one expression (using multiple columns) and variability in the step size of the independent variable. The user specifies the expression(s), the beginning and ending values of x, and the step size. To solve an equation (single variable) in this environment means to set up and manipulate a table having a column for each of the two expressions on either side of the equal sign. The actions are directed at determining values of x such that the two expressions are equal. Finding roots of functions is, of course, a special case. Despite the fact that little is novel here except the convenience of generating and modifying the data tables, the pedagogical consequences as described by Heid are far-reaching.

Here the table of data is changed from primarily a display representation to an action representation of the equality-of-functions statement inherent in an equation. Conversely, the equation is changed (at least temporarily) from an action to a display representation. Importantly, the actions on the table of data are primarily *semantic* in the sense that the student focuses on the numerical meaning of the actions: The activity of finding the x's which yield equal values in the other two columns of the table is guided by the values of the two expressions involved and not by syntactical considerations.

Algebraic coordinate-graph interpretation. Other ETC/EDC learning environments (Schwartz, Yerushalmy, & Harvey, 1988) and similar environments developed at WICAT (Lesh, Post, & Behr, 1987) and by Wenger, at the University of Delaware, approach the solving of single-variable equations by displaying the respective expressions on each side of the equation as graphs in *G*-2. In effect, they overlay the graphs of two functions, plotting Y_L = (left side expression) and Y_R = (right side expression) on the same coordinate axes. In an adjacent window appears the equation, which is subject to the usual algebraic equation-solving actions (add or multiply constants or variables on each side, etc.). The key to this environment is the hot link between actions on equations—objects in *E*-1—and the results of these actions in *G*-2. Here *E*-1 is the usual notation system with the usual syntax, but just as in the linking of expressions and graphs, *G*-2 plays a new role of representing *actions* on objects and not merely objects. This time the object, represented by two superimposed graphs, is an equation (an assertion about the equality of two expressions) rather than a single expression.

New ways of thinking about actions on equations and their consequences result. Now, when one adds or subtracts constants from both sides of an equation, a vertical displacement of *both* function graphs takes place, and the x-coordinate of the point of intersection of the two graphs does not change. Adding a linear term to both sides changes the slope of a line or the "shear-tilt" of a parabola—again, with no change in the x-coordinate.

However, if one mistakenly adds something to one side and subtracts it from the other, there is an immediate shift in the x-coordinate of the intersection, a change in the solution. This particular mistake, noted and studied by Kieran (1982), amounts to a subtle sign error that seems to be supported by a "redistribution scheme" based on the fair-share notion that if one subtracts from one number, one should add to its companion. This kind of reasoning leads to the following transformation rule applied to simple linear equations:

$$a + x = b \implies a + c + x = b - c$$

Except for the minus sign, of course, this is a correct transformation. Here is a good example of how the inert medium in which E-1 has historically been instantiated puts a large cognitive burden on the user to detect inappropriate actions on the basis of very subtle mismatches between patterns learned as correct and patterns produced in error.

Note that an equation solution in the form $x = c$ is displayed as the intersection of two lines in G-2: the diagonal $y = x$ and the horizontal line $y = c$. Thus, for example, if one were dealing with an equation having linear terms on both sides, the solution actions would take the form of "slides and tilts" directed at transforming the two given lines into the diagonal/horizontal form as indicated—all without changing the x-coordinate of their intersection.

One can also multiply or divide by linear terms. These actions allow for the introduction of extraneous roots and other peculiar, but interesting, phenomena whose mysteries are revealed in the graphical representation. For example, the introduction of extraneous roots to linear equations via multiplying by x is easily, and literally, *seen* to be a consequence of turning straight lines into parabolas, which may then intersect in more than one point.

Statements about equivalent equations and solution sets, based on syntactic manipulations of the notation system E-1 of single-variable equations, and guided by the abstract logic of conditionals, now gain new meaning and accessibility when represented as visible action consequences in G-2. Again, the ability of the computer to translate instantaneously across representations allows one to exhibit and hence *evaluate* one's actions in more than one representation system, without the cognitively prohibitive costs of performing the translation oneself.

Numerical coordinate-graph interpretation. Another single-variable equation-solving environment makes a more numerically oriented use of E-1 and G-2, together with an elementary action on the G-2 side. The action is simple: Move an x-cursor left and right (it appears as a dotted vertical line crossing the graphs of both the left and right sides of the equation). Two dashed horizontal lines project to the y-axis from the intersections of the vertical x-cursor line with the graphs of the respective sides of the equation. Here, solving an equation amounts to sliding the x-cursor, using either cursor movement keys or a mouse, to align the two horizontal projections. "Finding x" becomes a real search activity with a kinesthetic feel. One can vary the step size and rescale as necessary to zoom in on potential roots.

This system parallels the table-centered system of Fey and Heid described above. The basic difference is the degree of control and focus on particular x-values afforded by the graphics x-cursor versus the quantitative explicitness afforded by the table of data. Since the dynamic tables automatically order the data, the usual (static medium) difference between tables of data and coordinate graphs with regard to ordering no longer holds. If *x*-cursor movement is made discrete, then the coordinate graph has been rendered discrete, to complete the parallel with *T*-3. Thus we have an example of what might be termed *isomorphic* representation systems, as illustrated in Figure 5.

These kinds of modifications made to the basic notation schemes *T*-3 and *G*-2 make them almost parallel representations of *E*-1. Each supports a direct form of numerical-approximation equation solving. In fact, one can add a table-of-data window to the graphics environment so that all three representations are available simultaneously on the same computer screen, with different colors matching each side of an equation with its respective graph and table column.

One could, instead of overlaying graphs of the two sides of an equation, display the two graphs adjacently, one for each side of the equation, on identically scaled coordinate systems. The *x*-cursor movement would take place simultaneously in both coordinate systems and the resulting dashed horizontal lines would cross the *y*-axes of both coordinate systems. This type of display emphasizes the separate identity of the two sides of the equation as distinct functions.

In summary, we have seen three different meanings of the phrase "to solve an equation," depending on which representation the activity occurs in and how that activity is reflected in a second representation. The fact that there are at least two representations involved is the critical factor—the student is not merely pushing symbols in isolation—especially if the functions and/or the data originate as descriptions of meaningful phenomena, which is entirely realistic given the capacity to deal with numerical approximations that is inherent to the systems described above.

The Shortcutting of Numerical Experience

The classic *Green Globs*, (Dugdale, 1983) and its more flexible derivative *Algebra Arcade* (Wadsworth Electronic Publishing Company, 1986) offer

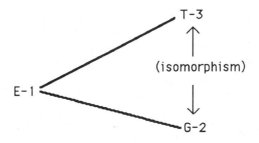

Figure 5. Isomorphism of three-column tables and two-dimensional graphs as representations of single-variable equations.

environments that involve the student in writing equations of functions whose graphs pass through given objects on a coordinate system. An interesting feature of these games and numerous other direct-translation graphing utilities is the frequent bypassing of the numerical aspect of functions. The student inputs the alphanumerical representation and gets the graphical representation in response, with no table of data, hence no explicit quantitative experience mediating the translation, as is the case when students plot functions on a point-by-point basis.

The cognitive consequences of this shortcut on students' conception of function are not clear. I have been investigating the role of numerical thinking in the translation process using a modified version of another computer game modeled after the W. W. Sawyer/Robert Davis activity *Guess My Rule* (Barclay, 1985). In this activity, the computer has a "secret" function (taken from a list built in advance) that the student is to guess (as an algebraic formula) on the basis of the computer's response to his/her numerical inputs. The student can *choose* the form of the computer's feedback—either numerical or graphical (the computer evaluates the student's guesses for each input as it is made). Early results of this work suggest that students in first- and second-year algebra courses are primarily arithmetic creatures (see also Booth, 1984) and that significant teaching and prompting are needed to get them to use graphically represented information, even when it is the most efficient means to a solution (Kaput, 1988a). Other classifications of the students into "algebraic" versus "pre-algebraic" categories are possible based on how they use the idea of variable in their reasoning processes.

From Functions to Families of Functions

An interesting option in the *Algebra Arcade* game allows a parametrically defined function family to be fixed so that the student plays with only that family and needs only to specify its parameters for each function input. Similarly, one can build lists of functions in *Guess My Rule* that systematically vary parameters of a given family of functions. These sorts of features, also available in many graphing utilities, help elevate student attention from individual functions to families of functions in *F*-1.

There are also software environments that support building a function to match a given function which is given graphically, but not alphanumerically. The student inputs or increments appropriate parameters, with the result of each entry graphed in a way that allows direct graphical comparison with the target function. A particularly revealing action that such a system supports is inputting a range of values and a step size for a given parameter, so that the student can observe by means of near-animation the role of a given coefficient as the system steps through the specified interval, graphing each function associated with the new value of that parameter (and deleting the previous function, if desired).

Note the difference between this kind of incrementing and the analogous activity in *T*-2 or *T*-3 described earlier. Here the process is shifted upward in abstraction: *Functions* are being incremented rather than their independent variables. Care needs to be taken, however, to distinguish variation in a function from variation in the parameters that define the function (Goldenberg, 1987), particularly with regard to the student's understanding of the entities being dealt with (Wagner, 1981). Nonetheless, shifting the object of attention from the individual function to parametrically defined *families* of functions could have significant curricular implications. Indeed, the complex web of ideas associated with the word *variable* may become more accessible, given the availability of a medium in which variation is dynamically possible, and where invariance under controlled variation can be systematically and visually investigated (Kaput, 1988b).

NEW ACTIONS ON OLD REPRESENTATIONS

A variety of different display representations have been used over the years to render aspects of algebra more learnable in static media. Some of these might be more effective in the computer medium, depending on whether or to what degree their representational function is enhanced by incorporating dynamic and/or interactive aspects, as indicated in the following examples.

Coordinate graphs. Except for equation-solving activity based on cursor movement, the above examples made little use of coordinate graphs as action representations. Even in the expression manipulation environment, *G*-2 was used to display actions initiated on polynomials rather than as a place to initiate actions. Graphs have traditionally served as display representations because the only readily available media were static. *But the static medium restriction is no longer in force!* With the advent of relatively inexpensive bit-mapped graphics displays, we are now free to move and manipulate graphical objects just as we have always been free to manipulate alphanumeric objects.

Perhaps the most obvious coordinate-graph-based action is simply to drag the graphs of equations or functions, with their changing equations displayed automatically. Limited versions of such software have been commercially available for several years and are newly available in a sophisticated collection of graphically-oriented algebra learning environments (Schwartz, Yerushalmy, & Harvey, 1988). This new type of software supports rotations, horizontal and vertical translations, and other appropriate actions on graphs of polynomials and other relatively well-behaved functions. One such action, which corresponds to dividing by selected linear terms, amounts to "unbendings," if the divisors are roots, and can be very revealing as a root-finding activity.

On the other hand, one can build single-variable polynomials using graphically-oriented transformations applied to objects in G-2. A typical activity is the building of a function equation to match a given graphically defined function. The student begins with a most basic of functions, the line $y = 1$, and performs graphically defined actions, such as up/down and left/right translations, as well as "tilts" (slope changes), "bends" (products by linear expressions), and vertical scale changes. The computer tracks the equation representation of the resulting function while performing the directed graphical actions. Bends are particularly interesting because of the roles and invariance of roots in the function-building process.

Area of rectangular geometric figures. For the ancients, a primary reference field for polynomial algebra was the display and measurement of rectangular geometric figures. Virtually all the properties of elementary algebraic manipulations can be modeled in these terms. It is not hard to imagine a split-screen microcomputer environment that uses actions on, say, rectangles in one window, with their formal polynomial analogs in the other, with the values of variables under user control via the lengths of sides.

Balance scales. Balance scales have often been used to emphasize the need to preserve equality when acting on equations, as well as to give a concrete representation of the idea of an unknown. Primitive interactive versions of this model are commercially available. One of several difficulties inherent in the balance-scale model is that "more" is represented as a downward movement, whereas in cups, thermometers, coordinate graphs, and other models, "more" is "up"—a good example of inherently conflicting representations (Booth, 1987; Kaput, 1987d).

Function machines. Function machines are good candidates for computer portrayal, especially if the structure of the function equation can be directly reflected in the layout of the machine. Versions of such have been produced by Kleiman (1986) for 8-bit microcomputers and in beta form for 32-bit computers by Feurzeig, Richards, and Carter (1988). Each uses a split-screen, linked representation approach, and each supports the building of machines from the four primary operations. The latter has the additional feature of a hierarchical display of the structure of a function, so that, for example, an expression of the form $(2x + 3)^4$ can first be represented as a "fourth-power machine," which can then be "exploded" to reveal a submachine that doubles and adds 3.

Inversion of functions, which is represented as a reflection across the diagonal in G-2, has a more procedural representation in the function machine. Active tracers, such as a dashed line growing from left to right in T-2 or from the x- to the y-axis via plotted points in G-2, demonstrate function inversion as a literal reversing of coordinates. This kind of dynamic representation also lends a temporal procedurality to a traditionally static representation and pinpoints the breakdown of the process when the function is not one-to-one.

NEW NOTATIONS AND NEW ACTIONS

The plasticity of the computer medium allows a high degree of freedom to create new notations, and its dynamic capability allows a high degree of freedom to create new actions on these notations. Let us consider some typical examples.

Parsing trees. Although the parsing tree has been used over the years to conceptualize the syntactic structure of equations and expressions, it has seldom been used in the teaching of algebra. Thompson (1987) has developed a prototype dual-representation learning environment that exhibits expressions and equations in two forms—in their usual alphanumeric format (as character strings) and as a parsing tree. One can then manipulate both by using actions chosen from a representation-independent menu (see Thompson, this volume, for details). This system provides exhaustive representation of the syntactic structure of algebraic objects and actions on them—a property not shared by tables of data, coordinate graphs, or even alphanumeric statements (because of the tacit treatment of order of operations, for example).

Note the difference between the tree representation of actions on expressions or equations and the parsing tree: The former displays the syntactic structure of actions on objects, whereas the latter displays the syntactic structure of the objects themselves. Thompson's environment has shown promise in eradicating certain syntactic error patterns and in enhancing a very general concept of variable as an algebraic placeholder. It is interesting that such a concept of variable is *not* numerical, but almost purely linguistic in character.

Graphical pointwise addition of functions. Many of us have had occasion, when trying to discuss vertical displacement of the graphs of functions and perhaps even the addition of simple functions, to draw vertical "toothpicks" above the graph of the initial function to represent the pointwise addition process. Mark Konvisser at EDC has recently developed a prototype program which provides dynamic toothpicks (he calls them and the program *Measuring Sticks*). Moreover, the program supports (directed) variable measuring sticks, so that, for example, given a graph of x^2, one can add x to it by sliding a growing measuring stick (of length x) along the graph, so that the upper point of the stick traces out a plot of $x^2 + x$. Since these vertical sticks are directed, they point downward to the left of the origin, and hence have the effect of pulling the left side of the parabola downward, thereby illustrating the not-well-understood phenomenon of shear that results from adding x. Moreover, if one goes on to examine the roots of the resulting sum, the relation among the roots of $(x)(x + 1)$, the original form $x^2 + x$, and the graph becomes quite clear.

NEW REPRESENTATIONS OF THE MODELING PROCESS

The examples discussed so far have involved relations among mathematical notations—one mathematical notation representing another—rather than the use of mathematics to model quantitative aspects of non-mathematical phenomena. Modeling environments are harder to build but are likely to become more numerous once their power is recognized.

The Algebraic Proposer. Schwartz (1987, 1988) has developed a rather complete quantitative modeling environment that supports the solution of one- and (with a bit of stretching) two-variable word problems. It incorporates the traditional three representations—numerical, graphical, and alphanumerical—together with some aspects of the numerical/graphical equation-solving process described earlier. While these features alone make it a rich problem-solving environment, the reason for including it in this section is a new representation it provides for the modeling process and the non-inferring tutorial help based on that representation.

One enters problem "givens" (complete with units) and variable(s) and then forms polynomial expressions using these. Finally, one equates expressions and solves the resulting equations. The computer automatically generates a binary graph whose nodes are the givens, variables, and expressions. The graph of whatever part of the problem is already done can be called up for viewing at any time. If such a problem has been done in advance (by the teacher, for example), a user can call up the graph for help purposes and view notes written by the original problem solver regarding any node constructed up to that point. The machine will also list all the quantities that can be constructed using the student's input or will suggest that other information must be provided. Setting up an equation amounts to tying two loose ends of the binary graph together—usually an expression and a variable. Once an equation is built, one can solve it in the graphical or tabular fashions described earlier.

The binary graph provides a whole new view of the semantic structure of word problem situations. Moreover, one can go on to modify the given assumptions of the problem in a variety of ways, redo the problem, and compare the binary graphs. In this way, one can begin with a very simple situation and systematically elaborate it to build more complicated models. The system can also graph the area or slope functions of any function that has been constructed, thus providing entry into more advanced ideas as well.

The last important feature relates to the procedure-capturing facility described in the next section. If the computer has been "taught" a solution for a particular set of problem givens, then it can immediately solve any problem with the same semantic structure. Moreover, it can solve the problem with any constant provided as a parameter and then plot the solution as a function of that parameter. One can then address the issue of sensitivity of the solution to different problem givens (quite interesting in compound interest and other exponential growth/decay problems, for example). The kinds of insights accessible as a result of this procedure-capturing feature are the *real* goals of the modeling process. To paraphrase the pioneer computer scientist R. W. Hamming, the goal of modeling is not numbers, but insight.

Word Problem Assistant. Thompson (1988, this volume) has developed another prototype algebraic modeler that is based on earlier software by Greeno (1985) and which provides yet another way of illuminating both the structure of the situations being modeled as well as the processes by which models are constructed. An intriguing feature of this highly interactive, graphics-based system is its ability to support the convenient parametrization of any or all of the givens so that the "answer" can be automatically redescribed in terms of the assigned parameters. The fence between arithmetic and algebra is thus bulldozed and replaced by an easily traversable path.

MathCars. A new genre of dynamic simulation computer software is likely to bridge even more completely the chasm between phenomena and their formal representation within mathematics. I have recently outlined the design of a motion-modeling environment that enables a student to lay out the route of vehicles in the plane and to specify their motion either in a traditional way (giving speed or displacement numerically, graphically, or alphanumerically), or in a less traditional way (dragging them around the screen using a mouse), or even in simulation mode (driving them using a mouse-controlled accelerator). On the control panel beneath the simulated windshield are available not only an odometer and speedometer, but also other traditional and formal ways of representing the vehicle's motion in "real" simulated time. A wide variety of activities is then possible—driving a given trip, where the motion is specified in any of the formal ways; simulating real trips, such as one's ride home on the school bus; or comparing the motion of two or more vehicles (e.g., who goes farther in a given amount of time, given two different speed graphs?). By adding a current or wind to the path of the vehicle(s), a whole new collection of challenging activities is possible.

By carefully designing a motion modeler, one can generalize it to model the change and accumulation of other continuous quantities or even to accept real-time input of data from actual measurements. Such simulation environments, while providing intimate connections between students' personal experience and formal mathematics, also blur the distinction between algebra and calculus, even the distinction between mathematics and science.

CAPTURING AND BUILDING PROCEDURES

Applications of symbol manipulation software in schools have been rather direct, in the sense of using it as a tool in approximately the same styles as it is used in practice—to relieve the user of the burden of various calculations and to free attention and energy for more important tasks, or to widen the range of feasible tasks. This situation will change as symbol manipulation systems specifically designed for educational purposes become available (Kaput, 1988b). Some of these new systems will enable the user to perform certain procedures on particular cases (say, solve a particular linear equation), while the system "captures" or records the sequence of actions as a more general procedure to be executed on other linear equations, much as *The Geometric Supposers* (Schwartz & Yerushalmy, 1985) record geometric constructions as procedures to be re-executed on command. The user can then try the captured procedure on other equations (perhaps linear combinations of squared terms), examine where it works and where it fails, and modify it appropriately.

These environments might include a collection of primitive procedures and assembly mechanisms, with which the user can build complex procedures, including procedures that can be named and take arguments—a feature that blurs the distinction between programming and using packaged software. Sample activities in this sort of environment might be to "build a procedure that will factor the following expressions" or "solve the following equations" or "add the following fractions." By systematically adjusting the objects to which such procedures are to be applied, one could develop an appreciation of the underlying syntactical structure of algebra in a way better suited to life in the tool-rich environment of the future than through the repetitive exercise sets of traditional syntactical algebra—which not only alienate most students, but do a poor job of teaching skills that machines are better suited to do anyway.

Drawing on the style of the Xerox PARC *Algebra Toolkit*, Richards, Feurzeig, and Carter (1988) have developed a program they call *The Algebra Workbench*, which stores the steps a user takes in doing a traditional algebra manipulation problem. This program then graphically displays these steps in a binary graph whose nodes are the intermediate forms of the algebraic object being manipulated and whose edges are labeled according to the actions taken. This yields a process-trace that provides an explicit historical record of what has been done, which can then be compared with the trace produced by some grade of "expert," including that of a plodding novice whom the students can take delight in improving upon.

While existing examples of such process-reifying software are limited to fairly narrow domains, they illustrate a potentially major development. They transform temporally transient *processes* (which are otherwise representationally and pedagogically subordinate to their *products*) into spatially organized and permanent objects. As such, the processes are newly visible, inspectable, and therefore improvable. They open curricular opportunities for reflection on thinking and learning that are otherwise difficult to obtain.

My hope is that the technique of objectifying processes, and therefore strategies, will extend beyond the syntactic manipulation of formal symbols to activity with more semantic content.

ENHANCING SYNTACTIC MANIPULATION ENVIRONMENTS

Several groups (e.g., McArthur, Burdorf, Ormseth, Robyn, & Stasz, 1988) are working on environments to assist students in learning the syntax of the traditional alphanumeric algebraic notation schemes. The ultimate goal of much of this work is eventually to provide "intelligent tutoring" for students involved in algebraic manipulation tasks. However, given the difficulty of accomplishing this in the context of a reasonably full algebra transformation environment, most implementations currently embody more modest assistance, often in the form of context-sensitive help and intelligent problem generation (whereby the computer infers a weakness or "bug" and produces problems designed to challenge it).

Such enhancements of the formal algebraic notation system enrich the feedback provided to the student regarding the adequacy or appropriateness of his/her transformations within that system or enrichments of it—rather than linking the transformations to another notation system or situation to be modeled. They provide feedback on syntactical elaborations that are rooted in the same terms as the initial algebraic representation; as such, these environments share the same dubious curricular objectives and referential vacuity as do traditional algebraic manipulation exercise sets.

REFLECTIONS

Many have noted the potential of computer-based learning environments to renew and enhance the empirical side of mathematics and the associated inductive forms of learning and reasoning. This potential is based first on the capability of the electronic medium to support *interaction* between user and image, interaction that supports the construction of cognitive representations in ways very different from interactions with a static, inert medium. The second new capability inheres in the role of the computer as a fully plastic representational medium that supports all manner of notations and linkages among them.

A comparison between the applications of the technology described above and the list of sources of meaning in mathematics given early in this paper suggests that appropriate use of technology can support new meaning building across algebra. There can be, and are likely to be, fundamental changes in what it means "to know algebra." There will also be changes in the forms that knowledge can take (e.g., changes in relationships between conceptual and procedural knowledge) because old procedures will take on new meaning in new notations, new procedures will be possible in old notations, and entirely new notations and procedures will be created.

The past is a very poor guide to the future of this medium in algebra and engenders limited debate based mainly on narrow skill-development concerns: For example, how much syntactic skill development should be preserved in the curriculum? Our analysis has treated the different types of computer use as separate and distinct, but in the future they are likely to be available in virtually any combination, including combinations with notations not yet dreamed of, yielding a much deeper use of the technology than has typically been the case so far—deeper in both the penetration of the subject matter and in the level of cognitive interaction with the student. Further, the dynamic nature of the medium supports dynamic changes in variable values that render the underlying ideas of variable and function more learnable, which should make them accessible to a younger population, and which in turn makes possible a much more gradual and extended algebra curriculum, beginning in the early grades.

Indeed, the traditional prequisite relationships between algebra and other mathematics, such as calculus, are likely to be altered. After all, one way to view much of the syntax of algebra is as an adaptive reponse by the masters to the high cost of computation. They needed fancy algebraic techniques to be able to "see" the behavior of functions and to understand various kinds of quantitative relationships. Now, with the cost of computation minimized, those constraints no longer hold, leaving us free to make fuller use of numerical and graphical techniques, techniques typically of greater simplicity and generality (e.g., "guess-and-check" approximation) than the manipulation-intensive techniques that dominate most curricula and most conceptions of algebra.

The task of the researcher, curriculum developer, and teacher in creating, analyzing, and evaluating new learning environments will call upon all the intellectual resources we can muster. But I feel safe in asserting that profound changes in the learning, meaning, and doing of algebra will result.

REFERENCES

Barclay, T. (1985). *Guess my rule* [Computer program]. Pleasantville, NY: HRM Software.

Booth, L. R. (1987). Equations revisited. In J. C. Bergeron, N. Herscovics, & C. Kieran (Eds.), *Proceedings of the Eleventh International Conference for the Psychology of Mathematics Education* (Vol. I, pp. 282-288). Montreal, Quebec, Canada: University of Montreal.

Davis, R. B. (1984). *Learning mathematics: The cognitive science approach to mathematics education*. Norwood, NJ: Ablex Publishing Corporation.

Dugdale, S. (1983). *Green globs: A microcomputer application for graphing* [Computer program]. Sunnydale, NY: Sunburst Communications.

Feurzeig, W., Richards, J., & Carter, R. (1988). *Visdom: A visual programming language* [Computer program]. Cambridge, MA: BBN Labs.

Fey, J. T. (1984). *Computing and mathematics: The impact on secondary school curricula*. Reston, VA: National Council of Teachers of Mathematics.

Filloy, E., & Rojano, T. (1985). Obstructions to the learning of elemental algebraic concepts and teaching strategies. In S. K. Damarin & M. Shelton (Eds.), *Proceedings of the Seventh Annual Meeting of PME-NA*. Columbus: Ohio State University.

Freudenthal, H. (1983). *Didactical phenomenology of mathematical structures*. Boston, MA: D. Reidel.

Goldenberg, E. P. (1987). Believing is seeing: How preconceptions influence the perception of graphs. In J. C. Bergeron, N. Herscovics, & C. Kieran (Eds.), *Proceedings of the Eleventh International Conference for the Psychology of Mathematics Education* (Vol. I, pp. 197-203). Montreal, Quebec, Canada: University of Montreal.

Greeno, J. (1983). Conceptual entities. In A. Stevens & D. Gentner (Eds.), *Mental models*. Hillsdale, NJ: Lawrence Erlbaum.

Greeno, J. (1985, April). *Advancing cognitive science through development of advanced instructional systems*. Paper presented at the annual meeting of the American Educational Research Association, Chicago, IL.

Heid, M. K., & Kunkle, D. (1988). Computer-generated tables: Tools for concept development in elementary algebra. In A. Coxford (Ed.), *The ideas of algebra, K-12* (1988 Yearbook, pp. 170-177). Reston, VA: National Council of Teachers of Mathematics.

Kaput, J. J. (1979). Mathematics and learning: Roots of epistemological status. In J. Clement & J. Lochhead (Eds.), *Cognitive process instruction* (pp. 289-303). Philadelphia, PA: Franklin Institute Press.

Kaput, J. J. (1985). Representation and problem solving: Methodological issues related to modeling. In E. A. Silver (Ed.), *Teaching and learning mathematical problem solving: Multiple research perspectives* (pp. 381-398). Hillsdale, NJ: Lawrence Erlbaum.

Kaput, J. J. (1986). Information technology and mathematics: Opening new representational windows. *Journal of Mathematical Behavior, 5*, 187-207.

Kaput, J. J. (1987a). Representation systems and mathematics. In C. Janvier (Ed.), *Problems of representation in the teaching and learning of mathematics* (pp. 19-26). Hillsdale, NJ: Lawrence Erlbaum.

Kaput, J. J. (1987b). Toward a theory of symbol use in mathematics. In C. Janvier (Ed.), *Problems of representation in the teaching and learning of mathematics* (pp. 159-195). Hillsdale, NJ: Lawrence Erlbaum.

Kaput, J. J. (1987c). *The role of information technology in the affective dimension of mathematical experience: Some preliminary notes*. Paper presented at the annual meeting of the American Educational Research Association, Washington, DC. (Also to appear in D. B. McLeod & V. M. Adams (Eds.), *Affect and mathematical problem solving: A new perspective*. New York: Springer-Verlag)

Kaput, J. J. (1987d). PME XI algebra papers: A representational framework. In J. C. Bergeron, N. Herscovics, & C. Kieran (Eds.), *Proceedings of the Eleventh International Conference for the Psychology of Mathematics Education* (Vol. I, pp. 345-354). Montreal, Quebec, Canada: University of Montreal.

Kaput, J. J. (1987e, September). *The cognitive foundations of modeling with intensive quantities*. Paper presented at the Second Annual Conference on the Teaching of Mathematical Modeling, Kassel, FRG.

Kaput, J. J. (1988a, April). *Translations from numerical and graphical to algebraic representations of elementary functions*. Paper presented at the annual meeting of the American Educational Research Association, New Orleans, LA.

Kaput, J. J. (1988b, April). *Applying technology in mathematics classrooms: It's time to get serious, time to define our own technological destiny*. Paper presented at the annual meeting of the American Educational Research Association, New Orleans, LA.

Kaput, J. J. (in press). Representationalism and constructivism. In E. von Glasersfeld (Ed.), *Constructivism in mathematics education*. Boston, MA: D. Reidel.

Kaput, J. J., & Gordon, L. (1987). *A concrete-to-abstract software ramp: Environments for learning multiplication, division, and intensive quantity* (Tech. Rep. No. 87-3). Cambridge, MA: Harvard Graduate School of Education, Educational Technology Center.

Kaput, J. J., Luke, C., Poholsky, J., & Sayer, A. (1986). *The role of representations in reasoning with intensive quantities* (Tech. Rep. No. 86-9). Cambridge, MA: Harvard Graduate School of Education, Educational Technology Center.

Kaput, J. J., Luke, C., Poholsky, J., & Sayer, A. (1987). Multiple representations and reasoning with intensive quantities. In J. C. Bergeron, N. Herscovics, & C. Kieran (Eds.), *Proceedings of the Eleventh International Conference for the Psychology of Mathematics Education* (Vol. II, pp. 289-295). Montreal, Quebec, Canada: University of Montreal.

Kieran, C. (1982, March). *The learning of algebra: A teaching experiment.* Paper presented at the annual meeting of the American Educational Research Association, New York. (ERIC Document Reproduction Service No. ED 216 884)

Kleiman, G. (1986). *The math path* [Computer program]. St. Louis, MO: Milliken.

Lesh, R., Post, T., & Behr, M. (1987). Dienes revisited: Multiple embodiments in computer environments. In I. Wirszup & R. Streit (Eds.), *Developments in school mathematics around the world* (pp. 647-680). Reston, VA: National Council of Teachers of Mathematics.

Matz, M. (1980). Towards a computational theory of algebraic competence. *Journal of Mathematical Behavior, 3*, 93-166.

McArthur, D., Burdorf, C., Ormseth, A., Robyn, A., & Stasz, C. (1988, January). *Multiple representations of mathematical reasoning* (Project Report to the NSF Advanced Technologies Program). Santa Monica, CA: Rand Corporation.

Richards, J., Feurzeig, W., & Carter, R. (1988). *The algebra workbench* [Computer program]. Cambridge, MA: BBN Labs.

Salomon, G. (1979). *Interaction of media, cognition and learning.* San Francisco, CA: Jossey-Bass.

Schwartz, J. L. (1987). The representation of function in *The Algebraic Proposer.* In J. C. Bergeron, N. Herscovics, & C. Kieran (Eds.), *Proceedings of the Eleventh International Conference for the Psychology of Mathematics Education* (Vol. I, pp. 235-240). Montreal, Quebec, Canada: University of Montreal.

Schwartz, J. L. (1988). *The algebraic proposer* [Computer program]. Hanover, NH: True BASIC, Inc.

Schwartz, J. L., & Yerushalmy, M. (1985). *The geometric supposers* [Computer programs]. Pleasantville, NY: Sunburst Communications.

Schwartz, J. L., Yerushalmy, M., & Harvey, W. (1988). *The algebraic supposers* [Computer programs]. Pleasantville, NY: Sunburst Communications.

Tall, D. O. (1985). Using computer graphics programs as generic organizers for the concept image of differentiation. In L. Streefland (Ed.), *Proceedings of the Ninth International Conference for the Psychology of Mathematics Education* (Vol. I, pp. 105-110). Noordwijkerhout, The Netherlands: State University of Utrecht.

Thompson, P. (1987). *Expressions* [Computer program]. Normal: Illinois State University, Department of Mathematical Sciences.

Thompson, P. (1988). *Word problem assistant* [Computer program]. Normal: Illinois State University, Department of Mathematical Sciences.

Wadsworth Electronic Publishing Company. (1986). *Algebra arcade* [Computer program]. Belmont, CA: Author.

Wagner, S. (1981). Conservation of equation and function under transformations of variable. *Journal for Research in Mathematics Education, 12*, 107-118.

ACKNOWLEDGMENT

This research was supported in part by the Office of Educational Research and Improvement under OERI Contract No. 400-83-0041 and by the National Science Foundation under NSF Award No. MDR-84-10316. Any opinions, findings, conclusions, or recommendations expressed herein are those of the author and do not necessarily reflect the views of either the Office of Educational Research and Improvement or the National Science Foundation.

Critical Issues in Current Representation System Theory
or
A Reaction to: "Linking Representations in the Symbol Systems of Algebra"[1]

David Kirshner
Department of Mathematics and Science Education
University of British Columbia

The dynamic linking of mathematical representations through computer circuitry is a new and promising development in mathematics education. Many programs that take advantage of this capability are currently available, and many more are under development. The theoretical framework that Kaput presents in his paper has evolved mainly pursuant to analysis and description of such programs. As his many examples demonstrate, the framework is of unquestioned value to analysis, discussion, and categorization of software environments. However, the question that we are asked to consider at this conference, and which is taken up in this critique, is whether that framework can also serve more broadly as a metalanguage for research into the psychology of algebra.

REPRESENTATION SYSTEMS

The basic notions presented in the framework are those of representation system and mental representation. A *representation system* consists of a mathematical system linked by a well-defined mapping to another mathematical system or to a nonmathematical situation. Considerable attention is paid in the paper to a technical vocabulary of representation systems. For example, distinctions are made between *symbol systems*, wherein one mathematical system is a representation of another, and representation systems involving a nonmathematical object *modeled* by a mathematical system.

A mathematical system together with its syntax may be considered apart from a mapping to another object, in which case it is a *notational system*. *Structural syntactic features* prescribe the well-formed entities of a notational system; *action rules* describe allowable actions on such well-formed entities. In the context of a given representation act, it may be useful to specify the *medium* in which a notational system is instantiated, the equivalence classes of marks or inscriptions that define its *characters*, and so forth.

A number of very nice ideas fall out of this technical treatment of representation systems. The motivating concept of the theory is a decentralized notion of mathematical *meaning* as a web of linked representations. This is

195

neatly captured in terms of compositions of mappings: A mathematical system maps into another mathematical system, called its reference field. This reference field usually turns out to be a notational system of another representation system mapping into yet another reference field, and so on. Kaput's description of the inherent nestedness—and incompleteness—of symbol reference within mathematics is a provocative image. The recursive notion of semantic equivalence is also elegant: Two objects are *semantically equivalent* if (a) they map into a single object (or syntactic variants of a single object) in the reference field or (b) they map into semantically equivalent objects in the reference field. (Unfortunately, no examples of this construct are offered.)

MENTAL REPRESENTATIONS

The fact that elegant formalisms emerge from the consideration of representation systems is not altogether surprising, since representation systems are comprised, for the most part, of mathematical systems and formal mappings between them. This explicit, technical analysis of representation systems is of unquestioned value for analyzing, classifying, and discussing software environments. But, whether the framework can more broadly serve cognitive research in algebra hinges on how well it can meet the cognitive concerns typical of current research.

Cognitive concerns are addressed through explicit incorporation of *mental representations* in the theoretical framework. For example, Kaput illustrates how mental representations shadow the accompanying mathematical representations in a representation system. In the simplest case, the evocation of a mental representation involves "deliberate, controlled matching activity," but this reading process is not the general case. The features of the mental representation do not necessarily correspond to the features of the mathematical representation. More explicitly, in Kaput's Figure 3, it can be seen that evocation of mental representations may evolve over intermediate cognitive states.

At this point it must be noted that, while the metalanguage provides an excellent framework for analysis of mathematical representations, it provides almost no structure to discuss the building or transforming of mental representations. The theory suggests where to look for mental representations, but not how to find them. This is especially clear in the introductory section where "four sources" of mathematical meaning are presented. Of these, only the fourth (not pursued in the Kaput paper) actually describes the character of a mental representation. For the remainder, as in the paper generally, Kaput is content to describe the location of mathematical meaning without providing clues as to how it might be characterized.

REFLECTIONS ON LINKING REPRESENTATIONS

In proposing the representation system framework—developed for analysis and classification of software—as a metalanguage for cognitive research with a "futurist orientation," there is an implicit message that the psychology of algebra should (or will) reorient itself around the development, evaluation, and refinement of such software and abandon the detailed cognitive descriptions it now pursues. This implicit commitment deserves an explicit response.

Mathematics education research serves mathematics education practise. The kinds of cognitive concerns addressed in research today are pursued primarily, not for the sake of describing mathematical thinking as an end in itself, but because of the belief that ultimately such descriptions are necessary for the furtherance of practise. As a consequence, dramatic improvements in meaningful mathematics learning could serve to justify a decreased emphasis on cognitive description.

Expectation that the software under development may effect such dramatic improvement would seem to be the basis for such an optimistic prognosis. Kaput shares with a large proportion of the research community the belief that the current mathematics curriculum is fundamentally misdirected in its approach to algebra:

> One of the problems to be addressed by algebra research is that of student alienation, which, I suggest, is the result of teaching algebra syntax instead of semantics. This problem is compounded by . . . the lack of linkages to other representations that might provide informative feedback on the appropriateness of actions taken. (Kaput, this volume, 5th paragraph)

From this analysis it is not unreasonable to predict that any significant move to referential teaching of algebra will produce dramatic improvements in learning. The representation-linking software currently under development certainly represents a significant move toward referential teaching of algebra. Thus, we may be entering a world in which meaningful mathematics is more routinely attained, a world in which less detailed understanding of cognitive structure is required of mathematics education research.

Unfortunately, I do not share this optimism for the coming generation of computer software, nor do I subscribe to the above diagnosis for the malaise in mathematics education. I believe that the human mind is uniquely fashioned to learn syntax as syntax and that current syntactic instruction fails, not because it is syntactic, but because research has not begun to fathom the depth of complexity and intricacy required of syntactic performance in algebra (Kirshner, 1987). I believe that the natural predisposition of the mind is to approach new, structured domains syntactically. Regretfully, I will not be surprised to find that the computer environments designed to promote meaningful conceptual development result first and foremost in learning about button pushing on computer keyboards.

FOOTNOTE

1. This paper was written in reaction to an earlier draft of the paper by Kaput (this volume). A full discussion of the theoretical framework alluded to herein can be found in: Kaput, J. J. (1987). Toward a theory of symbol use in mathematics. In C. Janvier (Ed.), *Problems of representation in the teaching and learning of mathematics* (pp. 159-195). Hillsdale, NJ: Lawrence Erlbaum.

REFERENCE

Kirshner, D. (1987). *The grammar of symbolic elementary algebra.* Unpublished doctoral dissertation, University of British Columbia, Vancouver.

School Algebra for the Year 2000

James T. Fey
Departments of Mathematics and Curriculum and Instruction
University of Maryland

Concepts, principles, and methods of algebra are the core of secondary school mathematics. As a prerequisite for study in every branch of mathematics, science, and technology, algebra is the first high school mathematics course. Most college-intending students spend at least one year beyond algebra developing the algebraic skills and understandings required by trigonometry, analytic geometry, calculus, and statistics.

Unfortunately, there is broad agreement that, despite several years of study, most students do not acquire expected levels of proficiency in algebra. Even those students who do become skilled in manipulation of symbolic expressions often are stymied when faced with a task requiring application of those skills in situations of realistic complexity. However, the emerging capabilities of calculators, computers, and other electronic information processing technology suggest revolutionary changes in the goals and teaching of algebra. Technology is reshaping conceptions of what we ought to teach and what we can teach.

COMPUTERS AND ALGEBRA

When used for analysis of numerical, graphic, or symbolic information, calculators and computers offer powerful new approaches to familiar problems and access to entirely new kinds of important algebraic problems and methods. The best way to illustrate this potential and its implications for school mathematics is to consider a typical problem situation and the computer-intensive approaches that can be applied:

> A business that plans to introduce a new product is interested in potential sales, revenue, costs, and profit and in the relationship of these outcome variables to prices and investment.

Students in traditional algebra courses might be able to answer a few simple and straightforward questions in this situation. Students with access to computer tools and with the understanding of broad algebraic concepts and strategies can produce much more impressive work.

Finding Mathematical Models

The first important step in a modeling task is finding relations among the key variables. In the problem solving of typical mathematics courses, this requires translation of precise written conditions into algebraic equations, inequalities, or function rules. The ability to make these translations is particularly useful where mathematical methods are applied to situations involving quantitative principles of science or well-known formulas from domains like geometry or the mathematics of finance. However, making these translations requires confident and mature understanding of variables and expressions and their uses, with the usual consequence that realistic problem-solving experiences are scarce and come late in standard curricula. Some very interesting work by Kaput and Schwartz at the Harvard Educational Technology Center and by Tinker at the Technical Education Research Centers (Cambridge, MA) is showing ways that intelligent computer tools can help students learn formal modeling principles and procedures.

The availability of other computer-based modeling tools permits extension of algebraic methods to quantitative situations in which no existing formula relates the variables. In many realistic problems, the important relations among quantitative variables are usually determined from empirical data. With a modest amount of data, students can determine reasonably good linear fits by plotting the data, drawing a line that fits the data pattern well, and calculating the equation of that line "by hand." For more complex data sets, computer curve-fitting utilities make this phase of modeling and problem solving an easy task.

Numerical Methods and Spreadsheets

When rules relating important variables have been established, computer-based numerical methods can be used to search for answers to questions that involve algebraic equations or inequalities, using a table of values generated by student programming or a more complex matrix of values produced in a spreadsheet. With computer help in calculations, such systematic search methods are very effective. Furthermore, they give students a valuable numerical intuition about the problem situation and permit answering, by successive approximation, optimization questions of complexity far beyond the usual secondary school algebra fare.

A typical computer-based problem-solving session might begin by finding a demand equation from data (see Figure 1) and then proceed to find the price that gives maximum revenue or profit by a systematic search through a spreadsheet matrix of values for the interrelated variables (see Figure 2).

Graphical Methods

When relations among several variables have been established, computer graphics can be used to display those relations simultaneously, giving visual insight into questions involving zeros and extreme points for functions. Graphics resolution has now become fine enough to allow surprisingly accurate estimation of coordinates of single points (see Figure 3).

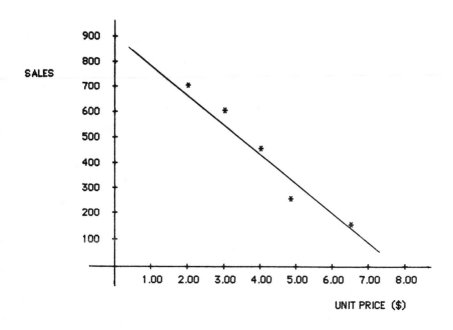

Figure 1. Best linear fit: Sales = − 125 * Price + 900.

PRICE ($)	SALES	REVENUE ($)	COSTS ($)	PROFIT ($)
2.00	650	1300.00	1475.00	-175.00
3.00	525	1575.00	1287.00	288.00
5.00	275	1375.00	911.00	464.00
8.00	-100	-800.00	347.00	-1147.00
6.00	150	900.00	723.00	177.00
4.00	400	1600.00	1099.00	501.00

Figure 2. Spreadsheet search for maximum profit.

Symbolic Methods

To provide exact answers for equation and optimization questions, computer symbol manipulation programs offer algebraic assistance much like the skills of an expert algebra student. Simple instructions like SOLVE, FACTOR, or SIMPLIFY produce formal transformations of algebraic expressions with numerical and literal terms. For example, in muMath a few simple instructions produce exact break-even points for the above business planning situation. The user need only enter the SALES function rule

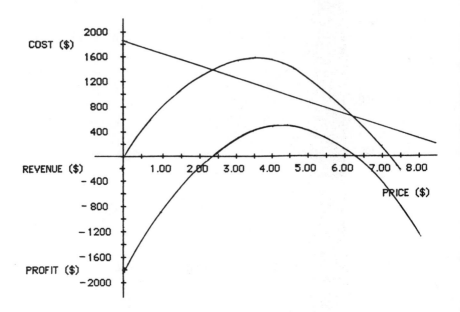

Figure 3. Cost, revenue, and profit as functions of price.

in natural algebraic syntax, define the REVENUE function as PRICE *
SALES, and ask muMath to solve the equation REVENUE = 0.

The striking and significant feature of these computer-intensive
approaches to algebra problems is that the user performs almost none of
the many symbolic procedures required by traditional approaches. In much
the same way that calculators take over the burden of algorithmic arithmetic
but still require the user to choose appropriate operations and enter appropriate numerical data, computers expect users to know what to do but not
the details of how to perform intricate manipulations. The computer user
must focus on global problem-solving strategies, or superprocedures, but
not on the execution of routine symbolic subprocedures required to solve
equations and produce graphs.

At the same time that computer tools offer powerful new approaches to
problem solving in algebra, they suggest fresh instructional approaches to
important algebraic concepts, principles, and skills. For example, computer
numerical, graphic, and symbol manipulation tools offer students multiple,
linked representations for the abstract concepts and relations embodied in
algebraic expressions. Each representation offers its distinct perspective on
the underlying mathematical ideas. Computer tools allow users to shift
quickly from one representation to another or to view several representa-

tions simultaneously, thus providing the opportunity and the challenge to choose the form that is most insightful.

The same symbol manipulation software that can be used to "get answers" for algebra questions can also be used to enrich student understanding of the fundamental concepts and rules that govern the transformation of algebraic expressions into useful equivalent forms. For instance, students can use muMath in a one-step-at-a-time mode to develop strategic skill in solving equations. The following sequence of questions for, and answers from, muMath shows how symbol manipulation software might be used in teaching about linear equations:

(User enters)	?	$35X - 459 == 306 + 18X;$
(MuMath responds)	@:	$-459 + 35X == 306 + 18X$
	?	@ + 459;
	@:	$35X == 765 + 18X$
	?	@ $- 18X;$
	@:	$17X == 765$
	?	@/17;
	@:	$X == 45$

Students can also use computer symbolic procedures to see patterns in the multiplication or factoring of polynomials. For example, the following sequence shows an intriguing result that is simply not accessible without some aid in the calculations:

(User enters)	?	$(X+1)*(X+2);$
(MuMath responds)	@:	$2 + 3X + X^2$
	?	$(X+1)*(X+2)*(X+3);$
	@:	$6 + 11X + 6X^2 + X^3$
	?	$(X+1)*(X+2)*(X+3)*(X+4);$
	@:	$24 + 50X + 35X^2 + 10X^3 + X^4$

In every case, using the computer to assist with technical details allows the user to focus on strategy—the connection between given information, chosen symbolic moves, and the results—without the distraction of executing, by hand, arithmetic or algebraic subprocedures.

With an exploratory tool of this power on hand, it seems possible that students can effectively discover and begin to use a variety of important general algebraic principles without following the traditional path through extensive prerequisite exercises in algebraic grammar. The computer tool also suggests a variety of ways to approach the teaching of traditional symbol manipulation procedures when students who are interested in or who will need that algebraic skill have gained a solid basis of understanding about variables, expressions, and equations.

CHANGES AHEAD FOR SCHOOL ALGEBRA

The preceding examples of computing at work illustrate only a few of the most obvious impending changes in the ways that people can learn algebra and use it to solve problems. In planning research that will improve algebra teaching and learning, it seems essential to study the subject as it will be ten or fifteen years in the future. Despite several decades of speculative discussion and writing on the prospects for technology, only exceptional schools have made computer-motivated changes in the goals or teaching of algebra. However, we have begun to see prototypes of hardware, software, and curriculum materials that realize the potential for dramatic change.

Hardware and Software Prospects

In thinking about the implications of computing for secondary school algebra curricula and teaching, the easiest part is assessing the technological prospects. It now seems modest to assume that in the very near future every mathematics student will have a personal notebook-sized computer with very large internal memory, dynamic high resolution graphics, and high speed numerical/logical/symbolic processing. Accompanying this hardware will be built-in software that includes programming languages and a variety of easy-to-use high level mathematical utilities, such as a spreadsheet that supports integrated numerical, graphic, and symbolic study of a collection of algebraic functions and relations. Furthermore, students will have access through networks to a rich library of mathematical, scientific, and technical data bases and information processing tools.

New Goals

The conceivable curricular responses to the changing environment for algebra range from modest enhancement of traditional syllabi to major restructuring of the entire curriculum. Algebra courses could remain substantially as they have been for years, with some more complex manipulations left to computers, or algebra could be integrated with topics from geometry, calculus, statistics, and science to produce a program based on mathematical modeling in which calculations are accomplished almost entirely by computer. The more radical changes offer hope for an impressive leap forward in the level of mathematical power acquired by secondary school students, but these changes entail risks. Nearly every suggestion for change is accompanied by cautions that research and development are needed to advise implementation decisions.

From my perspective, the proposed changes in the content of school algebra respond to five basic technology-driven challenges to mathematical tradition.

The skills/understanding connection. For many of us interested in technology and mathematics curricula, the most obvious implication of computer tool software is the opportunity to rebalance the relationship among skill, understanding, and problem-solving objectives in algebra. Over the past several decades, published syllabi for algebra courses have included a broad array of topics: sets, relations and functions, number systems, polynomials, rational expressions, exponents and radicals, logarithms, equations and inequalities, systems, sequences and series, conic sections and other topics from elementary analytic geometry, combinatorics and probability, trigonometry, and the basic structures of abstract algebra. This list describes a foundation of important conceptual and procedural knowledge that is invaluable in advanced mathematics and worthy of at least the normal two years of high school algebra allocated to it.

In fact, the agenda for high school algebra is overloaded with ideas and skills that prove difficult for many students. As a consequence, there are usually substantial differences among the intended, implemented, and achieved algebra curricula. Teachers and students must focus on some core of essential algebraic topics. When the College Board Mathematical Sciences Advisory Committee published *Academic Preparation in Mathematics* (1985), it listed the following essentials in algebra:

College entrants will need:
- Skill in solving equations and inequalities.
- Skill in operations with real numbers.
- Skill in simplifying algebraic expressions, including simple rational expressions.
- Familiarity with permutations, combinations, simple counting problems, and the binomial theorem.

Those students who will take advanced mathematics in college will need additional preparation in high school, including:
- Skill in solving trigonometric, exponential, and logarithmic equations.
- Skill in operations with complex numbers.
- Familiarity with arithmetic and geometric series and with proofs by mathematical induction.
- Familiarity with simple matrix operations and their relation to systems of linear equations. (p. 27)

This list, from a very influential testing panel, sends an unmistakable message to schools: The bottom line in algebra is training students in procedures for manipulating symbolic expressions.

The technology-based challenges to this skill-dominated conception of school algebra take several forms. One family of proposals suggests teaching students only those manipulative skills done more efficiently by hand than by machine or only those skills needed as a basis for the understanding and intelligent use of tools like function graphers or symbol manipulators. Implicit in these proposals is the hope that time saved by eliminating excessive practice of complicated skill routines can be reallocated to activities that build understanding of key algebraic concepts and problem-solving abilities.

The obvious difficulty in the "less skills" approach is deciding where to draw the line between manipulations to be assigned to computers and those to be done by the user. Any specific skill targeted for reduced emphasis in the curriculum will find defenders who illustrate its role in theoretical or practical science or in some aspect of advanced mathematics, particularly calculus (Coxford, 1985). Furthermore, many mathematics teachers are committed to the principle that proficiency in procedural skills is a prerequisite to understanding basic concepts and problem-solving strategies in any branch of mathematics. Despite these reasonable concerns about the effects of changing skill/understanding priorities, it seems inevitable that there will be a general drift in the direction of less emphasis on those manipulative skills that can be performed by computers and calculators. However, the interplay between conceptual and procedural knowledge and learning will continue to be an absolutely central question on which curriculum decisions can be advised by thoughtful research.

While acknowledging that computers will soon take over such important, but routine, algebra tasks as solving equations, graphing functions, or inverting matrices, Maurer (1983) has suggested another perspective from which symbol manipulation skills might retain their importance. Consider the algebra involved in mathematizing the following problem:

> An airline carries 1200 passengers per day between New York and Los Angeles at a fare of $325. Market research suggests that each $5 rise in the fare will cause a loss of 30 passengers. What fare will lead to maximum daily revenue?

Here, revenue is a function of price, and the conditions can be translated into algebraic form as follows:

$$R(p) = p[1200 - 30(p - 325)/5].$$

Though this expression matches the conditions, it might be far more efficient or informative to work with the equivalent forms:

$$R(p) = 3150p - 6p^2 \text{ or } R(p) = (-6p + 3150)p.$$

Thus, conventional algebraic skills may retain their importance as tools for transforming expressions into equivalent, illuminating forms.

On the other hand, many of the standard transformations, like factoring or simplifying or expanding into partial fractions, can also be performed by computer symbol manipulation, again raising doubts about the future importance of traditional manipulative skills. Surely even computer-aided users of algebra must have some sense of the equivalent forms available and what each has to offer. But perhaps this strategic knowledge can develop without prior mastery of all the component procedures. Again, the issue is interaction of conceptual and procedural knowledge. The unanswered question standing in the way of reducing the manipulative skills agenda of secondary school algebra is whether students can learn to plan and interpret manipulations of symbolic forms without being themselves proficient in the execution of those transformations.

Functions, relations, and applications. Much of the debate over computer impact in algebra has taken the traditional content and organization of the subject as given and has focused on the question, "How much skill is essential?" However, a few curriculum analysis and development projects are exploring fundamentally different ways of defining the appropriate content of school algebra (Fey, 1984).

There are many different reasons for including topics in algebra curricula and for requiring students to study the course. Algebra is a symbol system of unparalleled power for communicating quantitative information and relationships, it is a training ground in careful rule-governed reasoning, its development is a significant thread in the history of mathematics, and its theoretical structure is based on concepts and principles that generalize to provide organizing schema in nearly every other branch of mathematics.

Of all justifications for making algebra the core of secondary school mathematics, the most convincing is its contribution to problem solving in nearly every scientific discipline. Thus, in looking for curricular selection criteria, it seems especially important to analyze the ways that algebraic ideas and methods aid reasoning in various applied problem situations. Looking at the situations in which realistic applications of algebra arise, I have been attracted by several possible changes in approach to basic concepts of the subject.

For instance, effective matching of algebraic methods to important problems requires an understanding of *variable* that is richer and different from the common descriptions:

- a letter that represents a number, or
- a symbol that stands for any one of the elements of a given set.

I believe that there is power in being more faithful to the natural meaning of *variable* as a measurable quantity that changes as a situation changes. While those quantities are often represented in expressions by words, phrases, or individual letters, fundamentally a variable is a measured quantity.

Looking at the ways that variables arise in realistic problem situations, one is immediately struck by the fact that variables of interest nearly always occur in relation to others. The crucial questions are generally of the form, "What combination of input values leads to a specified combination of output values?" In many situations one or more input variables are used to predict one output variable—the output is a function of the input.

Of course, to make effective progress in the study of a functional relation among variables, it is essential to have some sort of algebraic rules expressing the relation and some means of transforming those expressions to derive new information. However, as we have illustrated earlier in this chapter, approaching these tasks with the array of computer tools now available leads one to a conception of algebra that is very different from the traditional package of symbol manipulation rules. Instead of focusing on the search for a single number that satisfies some equation with well-behaved integer coefficients, it is possible to look at the global pattern in the relation between variables, to study implications of changes in the problem parameters or goals, and to base all of this analysis on functional expressions derived from realistic data collection activities. Though conventional algebra courses are generally restricted to problems with at most two variables and two constraints, computer tools make it feasible to deal with more significant and interesting complexity.

In a computer-intensive approach to algebra as a study of functions and relations, students will be able to study very difficult problems in algebra and calculus without following the laborious regimen of skill building that is the ordinary preparation. What algebra students will need is a comfortable working knowledge of basic families of elementary functions and the use of various computer representations to study particular functions. For each major family of algebraic functions, the curriculum could cycle through a three-stage process of study:

> *Stage 1. Recognition and Mathematization*
> Through exploration of motivating situations, students should develop the ability to recognize types of functions and defining characteristics of each. Their mathematization skills should include the ability not only to translate rules or physical principles into function expressions, but also to use various computer aids for this process.

> *Stage 2. Answering Questions*
> Strategies for studying function behavior can progress from informal mental approximations to the use of sophisticated computer utilities, with a focus on knowing the calculations required by a question, selecting the most appropriate tools, and interpreting calculated results.

> *Stage 3. Formal Organization and Verification*
> Finally, students should arrive at a structured summary of principles and methods discovered in exploratory work, with appropriate deductive verification.

This pitch for organizing algebra around the study of functions and relations might seem like a thinly disguised variation on the less-manipulative-skill argument. On the contrary, it proposes development of manipulative skills only when a basis of concrete referents has been established and when the skills to be taught are of demonstrable, immediate value. Skeptics might question the feasibility of such a global approach by pointing out that the concepts of variable, function, and relation are not easy for students in current mathematics programs. Given this experience, it is reasonable to worry about prospects for a new computer-intensive program that depends so heavily on acquisition of those concepts. Furthermore, applications, at least in their traditional word-problem form, have always been a difficult part of algebra. Thus, it is possible that blending applications and algebraic concepts will pose an even more difficult learning task for unsophisticated students. Despite these reasonable doubts about the efficacy of a function-oriented/skills-later approach to algebra, the potential payoff in enhanced student problem-solving power and the promising results that are beginning to emerge from curriculum experiments make further exploration of this theme very important.

Organized complexity. While any new technology makes some previously difficult tasks very easy, the new-found power invariably escalates expectations. For instance, in the time before radar, computers, and earth satellites, most people were content with a weather forecast based on a look at the clouds, a wet finger held up to the wind, and a reading of the farmer's almanac. However, with current space science technology and computers that can run models with hundreds of thousands of variables, we have come to expect far more sophisticated and accurate predictions. While computer algebra tools make traditional skills seem trivial, they set a new level of expectation. Students must become much more adept at coping with complexity and the algebraic representations used to study it.

The branch of mathematics usually associated with multivariable problems is linear algebra. Real-world information is now frequently described in the language of vectors and matrices. Inferences from that information are based on computer-executed operations for which the primary user contribution is to formulate the problem, plan needed analyses, and interpret results. Current school curricula give very limited attention to the basic concepts and methods of linear algebra. At best, students are introduced to several methods for solving two or three linear equations in two or three unknowns, vectors in the coordinate plane, and the arithmetic of low dimension matrices viewed as arrays of numbers (not representations of transformations). Because this strand of the algebra curriculum has been designed for paper-and-pencil "technology," it seldom includes the notation, methods, or global structuring concepts that are needed in situations of realistic complexity (Ralston, 1985).

Computers offer some obvious and attractive opportunities to meet the challenge. Most computer languages provide powerful motivation for learning general notational schemes like subscripted variables. Programming or use of semi-programmed algorithms is a very promising approach to teaching and learning about efficient algorithms. Computer graphics provide visual displays of multivariate relations. The rapid calculations by computer programs for matrix operations, solution of linear systems, and linear programming provide an unprecedented opportunity for students to discover basic principles of linear algebra through exploration of test cases. Included in this last category is the very important modeling issue of sensitivity analysis—the study of the ways that changes in problem conditions or goals affect results.

In addition to the familiar methods of linear algebra, computer technology provides completely new tools for dealing with organized complexity. The variety of business and scientific spreadsheets, some with powerful underlying algebra processors, offer effective levers on multivariable problems in which the relations are not restricted to linear conditions. In quantitative reasoning tasks of the future, effective use of such new computer tools is likely to be of far greater importance than paper-and-pencil processing of algorithms which are limited to a few simple variables and relations. It seems essential to revise school curricula to treat these approaches to problem solving. Again, the fundamental question is how one can most effectively develop a comfortable working knowledge of such tools. Is the best approach a path through the traditional regimen of personal algebraic skills applied to simple cases? I doubt it.

Algorithms, recursion, and successive approximation. In any branch of science, one of the persistent tensions is the balance between experimental and theoretical approaches to problems. Although mathematics is in some sense a completely theoretical discipline, it has its own variation of the experimental/theoretical tension. Many mathematicians focus their attention on questions about existence and uniqueness of solutions to given problems, while others look for accurate and efficient procedures to produce the numbers, functions, or shapes promised by existence proofs.

One strong motivation for building the first modern computers was to find a tool that would produce numerical answers to practical computational questions. Though computers are now used to process many kinds of numerical, graphic, and symbolic information, the design of algorithmic procedures that construct specific answers to given questions has remained a fundamental theme in computer science. Whether the task is factoring a prime number, drawing a three-dimensional surface, or searching for information in a data base, the design of effective algorithms is a pervasive activity.

When computer scientists look at the school mathematics curriculum, particularly in algebra, they often propose new approaches to traditional problems—approaches that embody computationally accurate and efficient problem-solving methods. Furthermore, they urge greater attention to encouraging students to think in terms of designing algorithms when solving problems. There seems to be both a need and an opportunity to reexamine each topic in the current curriculum with an eye toward presenting computationally effective methods and developing a general algorithmic predisposition.

The implications of this computer-oriented approach to topics in algebra are diverse. For example, Engel (1983) has worked out plans for an algorithmic approach to much of secondary school mathematics. Others have suggested ways that individual topics can be treated from an algorithmic perspective. For instance, the Gauss-Jordan elimination method for solving systems of equations is usually favored over the traditional methods presented in secondary school algebra. Several methods for writing polynomials offer computational advantages over the traditional canonical forms, and several variations on the quadratic formula have advantages in limiting computational errors.

Within the broad theme of algorithmic approaches, there is a particularly interesting and important subtheme, *recursion*. Many of the most important numerical algorithms involve a structure that builds a result one step at a time, with the result at step $n + 1$ constructed from that at step n by a procedure in the heart of a program loop. Effective use of recursion requires the thought processes and formal methods of mathematical induction, a difficult topic not prominent in current secondary school algebra curricula.

Recursive definition of functions is particularly important in the theory of difference equations—a theory that, because of its compatibility with computer methods, is an increasingly common mathematical tool for problems throughout the sciences and technology. Topics as diverse as banking, biology, and the physics of motion are now being treated with models drawn from elementary difference equations. In many practical applications, these discrete methods threaten to supplant the conceptually more subtle continuous mathematics of calculus and differential equations. Discrete methods make very interesting problems accessible to students who have much less than the traditional preparation in mathematical skills.

There is also a close conceptual connection between recursion and an important family of computationally effective methods that might be described as successive approximation. Like the start of a formal recursive construction, most successive approximations begin with an initial value (often a guess), perform a test, and revise the approximation in some systematic way. With computers to speed the test phase, such methods become very effective. However, the strategy of guess-and-test contradicts some of the most deeply held beliefs about mathematics. For many people, mathematics—particularly algebra—is the subject in which there is always a systematic method for doing things, and guessing of any kind is not respected. Computer-based successive approximation relies on quite a different attitude, but it is an attitude with real problem-solving power. It seems worthy of inclusion in our approach to algebra. However, we know very little about the ease or difficulty of developing in students a facility for trial-

and-error methods. In fact, we know very little about the difficulties involved in the learning or teaching of algorithmic ideas and problem-solving methods of any kind.

Algebraic structures. Historical studies of algebra commonly trace development of the subject through two chronological and conceptual phases:

> (1) Early (elementary) algebra is the study of equations and methods for solving them.
> (2) Modern (abstract) algebra is the study of mathematical structures such as groups, rings, and fields—to mention only a few. (Baumgart, 1969, p. 234)

Although the modern or abstract facet of algebra figured prominently in the curriculum innovation of the 1960s and 1970s, it seems fair to say that only traces of that influence remain in the courses that most students take today. Algebra in secondary school is primarily the study of polynomial, rational, and exponential expressions over various subsets of the complex numbers—what is often described as generalized arithmetic. Nearly every branch of advanced mathematics is some hybrid of abstract algebra (algebraic geometry, algebraic topology, etc.). However, the structures and methods of abstract algebra have not been widely accepted as tools in secondary school mathematics.

Computer science presents an entirely new case for consideration of abstract algebra in secondary school curricula. As computers process numerical and symbolic information, they rely on methods and results from Boolean algebra, lattices, finite fields, and a variety of other algebraic structures not usually drawn from the real numbers. Thus, for students studying algebra in secondary school, it seems important that the subject be presented in a way that facilitates subsequent generalization to important abstract structures. Though it may not seem important to treat elementary algebra as anything more than generalized arithmetic, such a limited vision might, in fact, make the transition to broader views more difficult than if the underlying abstract structures were emphasized from the start.

In many ways, the algebraic problems in more abstract structures are much the same as those in elementary algebra. In Boolean algebra it is often very helpful to rewrite expressions in equivalent, more revealing forms. Information theory uses polynomial algebra over finite fields, with many of the important operations paralleling the familiar algebra of polynomials over integers or rational numbers. However, again, most of the routine algorithmic work in these new algebraic structures is done by machine—leaving the conceptualization and interpretation to the user. Training in specific manipulative skills of elementary or abstract algebra is of little long-term value. The recurrent fundamental question is how we can get students to acquire the kind of global perspective that is needed to guide and use the computational power at hand.

ALGEBRA CURRICULA AND THE RESEARCH AGENDA

Every recent analysis of curriculum, teaching, and achievement in secondary school mathematics has suggested that most students are not acquiring the mathematical abilities needed for dealing with important quantitative reasoning tasks. The concepts and methods of algebra are essential to successful work in nearly every branch of mathematics, but the algebra curriculum delivered to students in the future must provide intellectual power far beyond the standard course of today.

Aspects of current research on algebra teaching and learning promise very useful contributions. However, it seems clear that emerging technology will make dramatic changes in the goals of school algebra. In the process, the significance of several long-standing research themes will diminish, and new fundamental questions will arise to take their place. Prototype computer-intensive curricula in algebra are being crafted today through a combination of artistry and experimentation by imaginative teachers and curriculum developers. But the design and the evaluation of these new conceptions of school algebra can be enriched by insights from multiple research perspectives on the problems of teaching and learning that accompany the kinds of changes being envisioned for school algebra.

REFERENCES

Baumgart, J. K. (1969). The history of algebra. In *Historical topics for the mathematics classroom* (Thirty-first Yearbook, pp. 232-260). Washington, DC: National Council of Teachers of Mathematics.

College Board. (1985). *Academic preparation in mathematics*. New York: Author.

Coxford, A. (1985). School algebra: What is still fundamental and what is not? In C. R. Hirsch (Ed.), *The secondary school mathematics curriculum* (1985 Yearbook, pp. 53-64). Reston, VA: National Council of Teachers of Mathematics.

Engel, A. (1983). The impact of algorithms on mathematics teaching. In *Proceedings of the Fourth International Congress on Mathematical Education* (pp. 312-330). Boston, MA: Birkhauser.

Fey, J. T. (Ed.). (1984). *Computing and mathematics: The impact on secondary school curricula*. Reston, VA: National Council of Teachers of Mathematics.

Maurer, S. B. (1983). The effects of a new college mathematics curriculum on high school mathematics. In A. Ralston & G. Young (Eds.), *The future of college mathematics* (pp. 153-176). New York: Springer-Verlag.

Ralston, A. (1985). The really new college mathematics and its impact on the high school curriculum. In C. R. Hirsch (Ed.), *The secondary school mathematics curriculum* (1985 Yearbook, pp. 29-42). Reston, VA: National Council of Teachers of Mathematics.

ACKNOWLEDGMENT

This research was supported in part by the National Science Foundation under NSF Award No. DPE-84-70173. Any opinions, findings, conclusions, or recommendations expressed herein are those of the author and do not necessarily reflect the views of the National Science Foundation.

Toward Algebra in the Year 2000
or
A Reaction to: "School Algebra for the Year 2000"

Sharon L. Senk
Department of Mathematics
Syracuse University

The main thesis of Fey's paper is that "the emerging capabilities of calculators, computers, and other electronic information processing technology suggest revolutionary changes in the goals and teaching of algebra" (2nd paragraph). Fey develops this thesis in two parts. In the first part, he describes the capabilities of prototype software that can be used by secondary school students to develop algebraic concepts or to answer questions posed in the language of algebra. Specific and extensive references are made to spreadsheets, statistical packages, function-plotting software, and symbol manipulation programs. Programming and numerical methods are also discussed, but in less detail.

In the second part, Fey describes five broad sets of issues that this technology raises about school algebra. These issues are: the skills/understanding connection; functions, relations, and applications; organized complexity; algorithms, recursion, and successive approximations; and algebraic structures. Fey argues that, although some of these issues are not new, each is magnified or reshaped by the availability of powerful and inexpensive software. As he delineates these issues, Fey sketches a picture of what curriculum and instruction in algebra might look like in the future. From that sketch it is clear that Fey envisions an algebra course driven not just by technology but also by realistic applications, that is, by mathematics as it is used to solve problems in our world.

Overall, Fey develops his thesis clearly and thoroughly. In particular, his conceptual analysis of the potential impact of computer and calculator technology on what might be taught in algebra convinces the reader that charting the course toward the year 2000 is indeed a complex task.

On several occasions Fey reminds us that multiple research perspectives can advise and enhance the decision making of those who determine what and how algebra is taught. He identifies specific questions to investigate, most of which are about the impact of an applications-oriented, technologically-rich curriculum on students' learning. In general, these recommendations are well-conceived and worthy of follow-up.

However, Fey's views on the relative importance of various types of software are not clear. In an earlier work (Fey, 1984) he describes specific programs in BASIC or Logo that can help students learn fundamental skills or concepts in algebra, geometry, and calculus. That work influenced the

214

College Board (1985) to recommend that "every student entering college in the 1990s will need to be able to use computer programs to perform arithmetic and algebraic calculations" (p. 22), and every student expecting to major in science, engineering, or business will need the "ability to write computer programs to solve a variety of mathematical problems" (p. 23). Yet, in his paper in this volume, Fey makes no reference to specific programming languages, and the role of programming per se receives almost no attention. Has he changed his views? It would have been worthwhile for Fey to have commented on the College Board's recommendations specifically as they relate to algebra. He might also have given some guidance for research and development on how to balance the roles of programming and ready-made software as tools for learning algebra.

A more serious concern about Fey's paper is that it may leave the reader with the misleading impression that conceptual analyses and research on students' learning are sufficient to move us toward an improved school algebra curriculum. Recent research has consistently shown that educational change occurs in a socio-political context involving many people, products, and institutions (see, e.g., Goodlad, 1984; Leiberman & Miller, 1981; Moon, 1986). Unfortunately, Fey completely ignores four factors within the socio-political context that influence school algebra today and which, no doubt, will continue to influence school algebra in the future: teachers, textbooks, evaluation, and articulation. Unless the mathematics education community deals with these factors, reform efforts such as Fey describes will have limited impact on algebra by the year 2000.

TEACHERS AND COMPUTERS

Fey's vision of the teacher's role in the algebra of the future is not clear. He mentions teachers only twice in his entire paper. (In contrast, students are mentioned more than 30 times.) However, because Fey does not specifically advocate tutorial software, and because none of the material described in his paper appears to be self-teaching, the reader can only assume that he expects teachers to continue to play an active role in students' learning in the next century.

Becker (1987a, 1987b) reports that across subject areas and grade levels, the teachers most likely to use computers as part of regular instruction, as opposed to enrichment or remediation, are those who are most knowledgeable about computers. Unfortunately, at present most secondary mathematics teachers do not believe they know very much about computers. Weiss (1987) reports that more than 50% of mathematics teachers in Grades 7-12 consider themselves to be either somewhat unprepared or totally unprepared to use computers as an instructional tool. Only about 25% consider themselves to be well-prepared or very well-prepared.

Given that so many mathematics teachers feel unprepared to use computers, it is not surprising that most do not use them for instruction. Becker (1987a) reports that only about one third of mathematics teachers use computers in *any* instructional capacity during the school year and that most computer use by secondary mathematics teachers occurs in programming courses. Of all 9th-12th grade classes using computers in any subject, only about 5% of the computer use occurs in algebra or higher mathematics courses. Within these algebra and higher mathematics classes, almost 60% of the computer time is spent on drill or tutorials, and about 20% on programming. Thus, at most 1% of the computer use in Grades 9-12 is presently devoted to the kinds of activities Fey suggests. Given that two thirds of the mathematics teachers never use computers in instruction, the gap is enormous between the technologically rich algebra curriculum Fey describes and what is now implemented in schools.

This is not to imply that the school algebra program Fey outlines is unattainable, undesirable, or unworthy of further study. Quite the contrary! However, studies of the impact of technology on students' learning should take into account teachers' knowledge and beliefs about this issue. Any large-scale efforts to ensure curriculum change of the type Fey suggests should consider teachers' needs for education about instructional uses of computers. Ideally, we should address our teacher education efforts to both in-service and preservice institutions.

Weiss (1987) reports that secondary mathematics teachers generally have little formal coursework in applications of mathematics. Furthermore, the University of Chicago School Mathematics Project (Usiskin, 1986/1987) has consistently found that, although mathematics teachers are generally willing and interested in using realistic applications, few know where to find examples within the grasp of secondary school students (Hedges, Stodolsky, Mathison, & Flores, 1986). Thus, Fey's emphasis on realistic applications in the curriculum may also be a barrier to its implementation. This suggests that additional research and development need to address teachers' knowledge about applications of mathematics.

TEXTBOOKS AND TECHNOLOGY

Studies throughout this century have shown that, in addition to teachers, textbooks also have a powerful influence on what students learn about mathematics (Begle, 1973; Goodlad, 1984; Rugg & Clark, 1918). In fact, McKnight et al. (1987) argue that "in most U.S. schools, commercially published textbooks serve as the primary guides for curriculum and instruction; any significant reform effort must take this into account" (p. xviii). Becker (1987b) underscores the importance of textbooks by suggesting that technology be coordinated with existing curricular materials. Thus, in order to move toward the kind of school algebra Fey suggests, our research and development agenda must consider the production of textbooks that incorporate both realistic applications and technology in substantive ways.

Some efforts are already being made in this direction. Rubenstein et al. (1987) are currently developing a textbook on statistics and advanced topics in algebra that incorporates function-plotting software, a statistical package, and BASIC programming throughout the course. As part of the same project, McConnell et al. (1987) and I and my colleagues (Senk et al., 1987) are developing beginning and intermediate algebra courses that emphasize applications of mathematics and utilize scientific calculators in a central role, computers in a supporting role. Fey and Heid (1987) have begun to examine the extent to which beginning courses in algebra can effectively utilize software of the type Fey describes in his paper. Further research and development are needed to study not just the effects of such courses on students' learning but also the extent to which they are implementable by teachers.

EVALUATION AND EDUCATIONAL REFORM

Evaluation is also an important factor in educational reform. Students completing four years of high school mathematics of the type Fey describes are likely to have different algebra competencies than students who have had four years of traditional mathematics. They may be better prepared than students from traditional courses in using computers as a tool, in handling complex but realistic applications, and in expressing relations between algebra and other areas of mathematics; they may be less prepared than traditional students in performing symbol manipulations with paper and pencil.

For instance, it is quite possible that a student in Fey's algebra course of the future could solve the equation

$$-16t^2 + 96t + 196 = 0$$

by using symbol-manipulating software or by graphing the function

$$f(t) = -16t^2 + 96t + 196$$

and reading the values of t for which $f(t) = 0$. Such a student might even recognize that the function f describes the path of a projectile under certain physical conditions. But, the student may be unable to solve the quadratic equation with paper and pencil using factoring, completing the square, or the quadratic formula. How should this student's learning be assessed? How should the effectiveness of the algebra program be evaluated if large numbers of students end up with similar competencies?

On the one hand, when doing evaluation studies, we need to be fair to the new curriculum. Thus, techniques for assessment of individual learning, as well as program evaluation consistent with new curricular goals, should be developed and used. On the other hand, we need to take into account the role of algebra as a gatekeeper to the further study of mathematics, science, and other areas, particularly if students are moving from an innovative course into a more traditional one. Thus, traditional measures, such as standardized tests, may be appropriate for student assessment and program evaluation. How to reconcile these two directions for evaluation is not immediately clear. However, it is clear that, unless the relation between goals and evaluation instruments is addressed, the effectiveness of curricular innovations such as Fey suggests cannot be adequately assessed.

ARTICULATION AND EDUCATIONAL CHANGE

Articulation, which is related to evaluation, is also an important factor in educational change. Fey ignores questions like: To what extent will colleges and universities count such new high school courses as legitimate college preparatory work? After being admitted to college, where will such students be placed for further study of mathematics?

These are not simply hypothetical questions. Currently, many universities give placement exams that measure almost exclusively paper-and-pencil, manipulative algebraic skills. On such exams students are not allowed to use calculators or computers (see, e.g., the *Algebra Competency Exam*, Syracuse University, 1979, or the *Mathematics Test A/4B*, Mathematical Association of America, 1983). These exams are not at all consistent with the curriculum Fey envisions for the year 2000 and are likely to misrepresent what entering students know. Mathematics educators who believe that curriculum research of the type envisioned by Fey is appropriate need to encourage articulation between high school and college mathematics teachers. Students must not be caught between an innovative secondary curriculum, on the one hand, and a traditional college curriculum and placement program on the other.

SUMMARY

In his paper, Fey provides a clear description of prototype software and a careful conceptual analysis of some issues this technology raises about algebra curriculum and instruction. In so doing, he provides a visionary look at what school algebra might be in the year 2000 and suggests some research questions about students' learning in such an environment. However, in order for large-scale change to occur in school algebra, we must also examine the impact of socio-political factors on the mathematics curriculum. In particular, research and development regarding teachers, textbooks, evaluation, and articulation will complement the agenda outlined by Fey and help move us toward a more effective algebra program.

REFERENCES

Becker, H. J. (1987a). *Instructional uses of school computers* (No. 4). Baltimore, MD: Johns Hopkins University, Center for Social Organizations of Schools.

Becker, H. J. (1987b, February). Using computers for instruction. *Byte*, pp. 149-162.

Begle, E. G. (1973). Some lessons learned from SMSG. *Mathematics Teacher, 66*, 207-214.

College Board. (1985). *Academic preparation in mathematics: Teaching for transition from high school to college*. New York: College Entrance Examination Board.

Fey, J. T. (Ed.). (1984). *Computing and mathematics: The impact on secondary school curricula*. Reston, VA: National Council of Teachers of Mathematics.

Fey, J., & Heid, K. (1987, June). *Effects of computer-based curriculum in school algebra*. Paper presented at a meeting of National Science Foundation project directors, College Park, MD.

Goodlad, J. I. (1984). *A place called school*. New York: McGraw-Hill.

Hedges, L. V., Stodolsky, S. S., Mathison, S., & Flores, P. (1986). *Transition Mathematics field study* (Eval. Rep. No. 85/86-TM-2). Chicago, IL: University of Chicago School Mathematics Project, Department of Education.

Leiberman, A., & Miller, L. (1981). Supporting classroom change. In J. Price & J. D. Gawronski (Eds.), *Changing school mathematics: A responsive process* (pp. 52-64). Reston, VA: National Council of Teachers of Mathematics.

Mathematical Association of America. (1983). *Mathematics Test A/4B*. Washington, DC: Author.

McConnell, J., Brown, S., Eddins, S., Hackworth, M., Sachs, L., Woodward, E., Flanders, J., Hirschhorn, D., Hynes, C., Polonsky, L., & Usiskin, Z. (1987). *Algebra*. Chicago, IL: University of Chicago School Mathematics Project, Department of Education.

McKnight, C. C., Crosswhite, F. J., Dossey, J. A., Kifer, E., Swafford, J. O., Travers, K. J., & Cooney, T. J. (1987). *The underachieving curriculum*. Champaign, IL: Stipes Publishing Company.

Moon, B. (1986). *The "new maths" curriculum controversy*. London: Falmer Press.

Rubenstein, R., Schultz, J., Hackworth, M., Flanders, J., Kissane, B., & Usiskin, Z. (1987). *Functions and statistics with computers* (2nd pilot ed.). Chicago, IL: University of Chicago School Mathematics Project, Department of Education.

Rugg, H. R., & Clark, J. R. (1918). *Scientific method in the reconstruction of ninth-grade mathematics*. Chicago, IL: University of Chicago Press.

Senk, S., Thompson, D., Viktora, S., Rubenstein, R., Halvorson, J., Flanders, J., Jakucyn, N., Pillsbury, G., & Usiskin, Z. (1987). *Advanced Algebra* (field study ed.). Chicago, IL: University of Chicago School Mathematics Project, Department of Education.

Syracuse University. (1979). *Algebra Competency Exam*. Syracuse, NY: Syracuse University, Department of Mathematics.

Usiskin, Z. (1986/1987). The UCSMP: Translating grades 7-12 mathematics recommendations into reality. *Educational Leadership, 44*, 30-35.

Weiss, I. R. (1987). *Report of the 1985-86 National Survey of Science and Mathematics Education* (Report No. RTI/2938/00-FR). Research Triangle Park, NC: Research Triangle Institute.

Part II

A Research Agenda

An Agenda for Research on the Learning and Teaching of Algebra

Sigrid Wagner
Department of Mathematics Education
University of Georgia

Carolyn Kieran
Département de Mathématiques et d'Informatique
Université du Québec à Montréal

One of the goals of the Research Agenda Conference on the Learning and Teaching of Algebra was to develop an agenda for future research. Toward this end, small-group sessions were scheduled during the conference for the purpose of generating questions that should be researched in order to advance our knowledge of the learning and teaching of algebra. A draft of the agenda was circulated toward the end of the conference, and additional items were suggested by participants in oral and written comments. A few more items were added by the editors to reflect points made in the conference papers and large-group discussions. The agenda presented here is thus an edited and elaborated version of the working group's thinking up to 1988. We hope that the questions listed below will stimulate some attempts at answers; we are confident that they will, at the very least, stimulate further questions.

Before looking at the agenda itself, it will be helpful to consider some of the major themes that emanated from the conference. To a large extent, these are the same themes that run throughout the current algebra research literature and which, naturally enough, underlie most of the agenda for future research. In the section that follows, we present one possible framework for organizing ideas from the literature review papers in Part I of this volume. We have also tried to summarize in very general terms our current state of knowledge on a few research topics, but we have in no way done justice to the wealth of information and variety of perspectives reflected at the conference. We leave it to the reader to consult the conference papers and other primary resources for further details.

MAJOR THEMES

It was evident from the conference discussions that the question, "What is algebra?" has many different answers. Algebra plays a number of important roles in the curriculum. Yet, most of the research done to date seems to come from three major perspectives on algebra—algebra as generalized arithmetic, algebra as a representation system, and algebra as a set of rules.

Algebra as Generalized Arithmetic

Algebra can be viewed as generalized arithmetic in two different ways. On the one hand, we can think of algebra as generalized *arithmetic*, in which case we focus on the link between algebra and its numerical referents. On the other hand, we can think of algebra as *generalized* arithmetic, in which case we focus on the structural aspects of the number system.

Most research on the understanding and construction of algebraic concepts and principles has been done from the standpoint of algebra as generalized *arithmetic*. Usually the focus on numerical referents is quite explicit, whereas the regard for structure is more often implicit and may even be left to the reader to infer (see Kieran, this volume). Qualitative methods of data collection, such as clinical interviews and teaching experiments, have proven more useful than the traditional quantitative methods in probing students' concepts of variable, equation, and function, as well as their understanding of procedures for solving equations and finding equivalent expressions. Research results generally confirm teachers' observations about what students can do and what they do not understand, but most research goes beyond the level of classroom observation to reveal specific and often surprising points of confusion. The best research offers tentative theoretical explanations for the perceived difficulties.

Like all education research, mathematics education research tends to be more fragmented and ambiguous than we might like it to be and, as a focused field of inquiry, it is still in its infancy (see Kieran & Wagner, this volume). Nevertheless, there are some general patterns of findings beginning to emerge, especially from the research that views algebra as generalized arithmetic. It is fairly clear, for example, that numerical referents both help and hinder students in learning algebra. Many students can initially make sense of literal symbols only in reference to numerical values but at the same time may err in regarding these values as fixed or restricted in some (irrelevant) way. Further, the fact that arithmetic and algebraic notation schemes follow different conventions (in concatenation, for instance) may create cognitive obstacles for students (see Herscovics, Kieran, this volume), especially because both notational systems are used in algebra.

The strong emphasis in arithmetic on finding numerical results of indicated operations contributes to several kinds of problems in algebra. First of all, it reinforces an operational rather than relational interpretation of the equals sign, which students are slow to outgrow (see Kieran, this volume). Second, it leads to discomfort with expressions like $x + 3$ as answers to exercises—a discomfort that the initial emphasis in algebra on finding specific numerical solutions to equations does little to dispel. Third, it undoubtedly contributes to students' propensity for "simplifying" algebraic expressions by combining whatever numbers are available ($11x + 5 = 16x$) and performing operations just once ($2[3x + 5] = 6x + 5$), as they would in arithmetic.

Awareness of these kinds of difficulties can sensitize curriculum developers and teachers to adjustments in both the pre-algebra and algebra curricula that may assist students in making the transition from arithmetic to algebra. Explicit instruction has been found to be helpful in remediating many of the errors mentioned above. In contrast, some difficulties mentioned below seem fairly resilient, even in the face of concerted instructional efforts.

Algebra as a Representation System

Research that focuses on representational aspects of algebra has revealed some striking similarities in the difficulties that students have across different representation systems. For example, there seems to be a parallel between the ways that students misinterpret the alphanumerical representation system and the graphical representation system—a parallel that suggests that the difficulty may be attributable as much to cognitive development as to instruction. That is, when first dealing with literal symbols, some students have trouble going beyond surface associations to the next level of representation. These students, despite instruction, cannot get past the superficial notion that the letter s, say, stands for *students* (as an initial) to the deeper idea that s is chosen mnemonically to stand for the *number* of students. Similarly, in learning to interpret graphs of functions, some students cannot get past their initial thought that the graph of a velocity function is a picture of the actual motion, to the deeper understanding that the motion must be deduced from the graph (see Herscovics, this volume). Of course, neither of these distinctions is all that transparent—both involve a visual distractor and both are subtle enough to catch any of us off guard occasionally. But the fact that some students continue to misinterpret these algebraic representations, despite careful instruction and despite other evidence of understanding the problem contexts, suggests that there may be more going on here than meets the eye, so to speak.

Another idea that seems difficult for students to apprehend fully is the notion that a single symbol can represent many quantities at the same time. In the case of variables, students work first with (single-valued) unknowns and with expressions, in which only one value at a time can be substituted, so it is not surprising that the leap to relational variables is cognitively exactly that—a leap, not a small step. It is perhaps less clear why a similar difficulty should arise in the context of scaling, whereby students are initially either unable or reluctant to allow lengths of different objects to be represented along the same line (see Herscovics, this volume). What seems very clear is that there are aspects of algebraic representation that involve ideas or ways of thinking that do not immediately transfer from natural language or other everyday experiences.

There are two large areas of research that focus especially on representational aspects in algebra. One area is the work on word problem solving, in which linguistic issues related to comprehension and translation are central. Although relatively little word-problem-solving research draws explicitly on psycholinguistic theories, the findings suggest that the process of translating verbal problems into alphanumerical statements is considerably more complex than translating verbal statements from one natural language to another.

Nevertheless, many students rely on a direct, syntactic approach to solving word problems. These students tend to focus on (and remember) the superficial aspects of problems and often fail to recognize structural similarities between problems that differ in context. Other students are able to take a more meaningful, semantic approach to translating verbal problems into algebraic language. These students are more likely to recognize similarities in structure, especially among problems in familiar contexts (see Chaiklin, this volume). The fact that problem comprehension plays such a significant role in the solving of algebra word problems suggests that theories in reading and language might be helpful in future investigations in this area.

The second area of research activity in which the central focus is on representation is in the development of computer software. Although the interactive facility of computers makes issues of communication at least as important with computers as they are with textbooks, these issues have thus far been largely ignored. Instead, most development efforts have been directed towards capitalizing on the dynamic, spatial representation facility of computers. Issues related to visual representation—especially issues of multiple, linked representations—are just beginning to be researched in connection with current software development projects (see Fey, Kaput, Thompson, this volume).

Algebra as a Set of Rules

The third major research perspective in algebra focuses on algebra as a set of rules. Much of the early equation-solving and error analysis research in algebra was conducted from this perspective. More recently, this perspective is reflected in information processing research and in the work on intelligent tutoring systems (ITSs). Information processing research seeks to understand human reasoning processes by building rule-based computer models that mimic learner behavior (see Larkin, this volume). This work requires detailed analysis of tasks and solution paths. ITS research also uses task and solution analysis, together with the interactive facility of computers, to mimic expert teacher behavior (see Lewis, Thompson, this volume).

Though both ITS research and non-tutorial, linked representation software take advantage of the interactive feature of computers, these two approaches contrast sharply in the role accorded the student: In ITS research the student is regarded as essentially a passive learner—the burden is on the system programmer to anticipate any contingent behavior on the part of the student and to program a reasoned response on the part of the tutor. In linked representation software the student is regarded as an active constructor of mathematical ideas—the computer is a tool that assists the student in searching for patterns, generalizations, and solutions to problems.

THE RESEARCH AGENDA

The major themes identified above reflect some fundamental aspects of algebra and may be helpful in structuring the overall thrust of algebra research. But, these themes are too few and too broad to provide an effective format for presenting the dozens of specific agenda items that arose at the conference. Instead, we have organized the agenda around a number of traditional, mostly generic research perspectives—content, learning, teaching, algebraic thinking, affect, representation, technology, curriculum development, testing, and teacher education.

Many of the agenda items listed below were raised more than once during the conference and from more than one perspective. Although we have tried to eliminate unnecessary duplication, some questions do appear more than once under slightly different guises. This occasional redundancy serves to emphasize that the indicated headings are not categories of questions to be addressed but, rather, are perspectives from which to view research issues in algebra. With each item we have tried to indicate the tenor of the discussion that led up to the question. Sometimes the discussion reflected the current state of research in algebra; more often the discussion reflected general concerns related to the learning and teaching of algebra.

Some of the agenda items are of an empirical nature; others are more theoretically oriented. That is, some of the items are framed as researchable questions; others call for definitions of terms, identification of variables, or consideration of factors. Ultimately, the goal of all of the agenda items is to improve our understanding of how students learn algebra and thereby to improve our classroom teaching of algebra. We hope the agenda will be useful to all persons associated with school algebra—researchers, teachers, teacher educators, administrators, supervisors, curriculum developers, evaluators, and policy makers.

Content

Because so much research has focused on students' understanding of key algebraic concepts and principles—involving variables, expressions, equivalence, equations, and functions—it is not surprising that several issues discussed at the conference came from a content perspective. At the very outset, the conference organizers asked all speakers to address some basic terminology in their papers:

- What is *algebra*?
    - a. What is algebra, as it is currently taught in schools?
    - b. What should algebra be, particularly in view of continuing technological developments?

In addition, one working session of the conference was devoted explicitly to the consideration of what algebra is, at present, and what it should become. No clear and succinct definition of algebra was ever agreed upon; in fact, there was some sentiment for leaving it an undefined term.

Nevertheless, there was tacit agreement among conference participants that the use of literal symbols is generally perceived as being a characteristic of algebra. With the use of literal symbols comes a convenient way of representing arbitrary names and labels, unknowns, equations, polynomials, functions, and other abstract structures. Thus it is that algebraic ideas pervade the mathematics curriculum both vertically (from the earliest grade on) and horizontally (across every branch of mathematics). It is important, then, to consider algebra not only as a subject in itself but also in its relationship to other areas of mathematics:

- What are key *algebraic concepts*?
    - a. Are there unifying concepts that should be used to structure the algebra curriculum?
    - b. Are there algebraic concepts that can provide unity throughout the school mathematics curriculum?

There are also procedures that are typically identified with algebra, such as simplifying or evaluating expressions, solving equations or inequalities, and graphing functions:

- What constitutes an *algebraic principle* or procedure?
    - a. Are algebraic principles nothing more than generalized arithmetic principles, or are there algebraic principles that are fundamentally different from arithmetic principles?
    - b. Are there algorithmic procedures that are algebraic, as opposed to arithmetic or geometric?

Universal access to four-function, hand-held calculators has sparked a continuing debate over how much and what kind of skill practice is necessary in arithmetic. Efforts have been underway for some time to change the public perception of what basic skills are in the elementary grades. Now there are hand-held computers that can perform most of the symbolic manipulations traditionally associated with algebra. So, we have a parallel question at the secondary level:

- What are the *basic skills* of algebra?

The issue of defining the basic skills of algebra in a computer-rich environment raises a host of related questions that will be addressed in the Learning section.

Word problems have always been a significant, some would say notorious, component of any algebra course. With a shift in emphasis from procedures to problem solving, word problems will constitute an even larger proportion of the algebra curriculum. The growing interest in problem solving, together with pleas for more realistic, less contrived problems, makes it important to ask:

- What is an *algebraic word problem*?
  a. Are there word problems that are intrinsically algebraic rather than arithmetic?
  b. What makes a method of solving a word problem algebraic rather than arithmetic?

Learning

The fact that so many students have difficulty mastering algebra, and some never learn it at all, makes the learner perspective crucial in algebra research. Significant progress is apparent here, especially with the recent advent of interdisciplinary research teams of cognitive psychologists and mathematics educators working together to chart the cognitive processes involved in the learning of algebra.

A first consideration in research on learning is the definition of terms. Even if we had a clear definition of algebra from the content perspective, we would still need to answer separately the question of what constitutes "learning algebra":

- What exactly do students learn when they "*learn algebra*"?
  a. What are students' intuitive, pre-instructional ideas about various algebraic concepts?
  b. What characterizes students' post-instructional attainment of conceptual and procedural knowledge?
  c. What characterizes experts' knowledge of algebra? What do they know and what are they able to do?

Mapping the progression of students' conceptual and procedural knowledge as they grow from novice to expert in algebra may reveal predictable patterns in the acquisition of ideas. Delineation of levels or types of understanding is a common feature of general learning theories and would surely be useful in the area of algebra:

- What are the *levels of understanding* in algebra with respect to specific concepts/processes (e.g., variable, function, polynomial, equivalence, etc.)?

  a. How can we characterize/define these levels (e.g., intuitive, operational, relational, formal, etc.)?

  b. Are there cognitive hierarchies with respect to:

    —modes of representation (natural language, graphical, numerical, symbolic, etc.),

    —types of organizing schemata, and

    —other features, such as discrete versus continuous (functions) or one versus two variables (equations)?

  c. What are the mechanisms that underlie transition from one level of understanding to another?

  d. Is there evidence of décalage in algebra, that is, predictable variability in the developmental level of understanding of various topics?

  e. Are there constraints on the rate of development through the identified levels of understanding due to general stages of cognitive development?

Inherent in the analysis of what students learn as they learn algebra is the identification of misconceptions. A fair number of common misconceptions already appear in the literature, but their frequency and variety suggest that there are many more to be found:

- What are *common misconceptions* that students acquire in algebra and how do these misconceptions develop?

Misconceptions often develop from an incomplete treatment of a topic. Students fill in the gaps with false generalizations, some of which eventually get corrected, while others remain unidentified until a novel situation elicits a revealing response. Students who never encounter a sufficiently novel situation may continue to harbor certain misconceptions, yet be able to perform satisfactorily in a limited variety of circumstances.

More troublesome than the incomplete concepts are those concepts that in algebra run counter to ideas and conventions learned earlier in the mathematics curriculum. Changes in notation and meaning can present cognitive obstacles to students, who must unlearn certain familiar ideas and re-learn new conventions in moving from arithmetic into algebra:

- What are the major *cognitive obstacles* inherent in the learning of algebraic concepts and procedures?

  a. Which of these cognitive obstacles can be attributed to the learner's prior arithmetic/geometric knowledge?

  b. Which of these cognitive obstacles can be attributed to algebraic symbolism, notation, or linguistic structure?

  c. How much misuse of algebraic language can be attributed to influences of natural language?

Technological developments have raised the possibility of new and essentially different approaches to the classroom learning and teaching of algebra. At the same time, the availability of hand-held symbol manipulators capable of performing most of the polynomial/equation/ function operations of high school algebra raises the same kinds of questions about the learning and teaching of algebra as the four-function calculator does in arithmetic:

- What are the *implications of technology* for the learning of algebra?

  a. What is the relationship between symbol manipulation experience and conceptual understanding in algebra?

  b. What knowledge of algebraic structures and procedures is required to support algebraic problem solving in a manipulation-freed environment?

Because algebra, like the rest of mathematics, can be considered a written, rather than oral, means of communication, the literal and graphical systems of representation are of particular significance as visual symbol systems. The vast psychological literature on perception may thus be relevant to research on the learning of algebra:

- What is the *role of perception* in:

  —working with algebraic notation,

  —understanding algebraic equivalence, and

  —interpreting graphs?

Students' ability to translate between symbol systems may also affect understanding, and conversely:

- What is the relationship between *procedural knowledge* in one representational system and *conceptual knowledge* in that, or any other, representational system?

There are many general issues related to learning that are of specific interest in algebra. For example, the increasing importance of problem solving prompted conference participants to ask:

- What is *insightful learning* in algebraic contexts?

    a. Are there learner variables (facility with arithmetic, background in geometry, etc.) that are significantly related to insightful learning?

    b. Are there teacher variables (focus on problem solving, beliefs about the nature of mathematics, etc.) that contribute to insightful learning?

One undercurrent that ran throughout the conference was the question of a unified theoretical framework for research in algebra. One small-group working session focused on theories that have guided the research in algebra up to now, and one of the two closing plenary sessions dealt with theoretical issues. It was duly noted that different theoretical perspectives are useful for finding answers to different kinds of questions. Significant progress has been made in algebra learning research both by mathematics educators who tend to focus on the learning of specific content and by cognitive psychologists who study the learning process in a more general sense. The question that persisted was:

- Can we find a comprehensive and coherent *theoretical framework* that links the major research perspectives in algebra?

Whether or not we succeed in developing a theoretical framework that is particularly suited to research in algebra, the quest for such a theory may itself be instructive. To begin with, we need to develop theories of learning that have adequate complexity and specificity to explain learning processes in algebra. One way to do this is to refine general cognitive models of learning so that they apply to algebra:

- How can general *theories of learning* be elaborated so that they are more applicable to algebra?

Teaching

Many of the issues related to the teaching of algebra are issues related to learning, viewed from the opposite side of the lens. The advantages and disadvantages of different instructional techniques are usually the first questions that come to mind in considering research on teaching:

- What are the effects of *various pedagogical approaches* (e.g., function-based, technology-centered, technology-supported, problem-solving, situational, rule-oriented, structural, intuitive, etc.) on the understanding of different topics in algebra?

    a. How effective are new modes of instruction in algebra (e.g., worked examples, numerical methods, technology environments)?

b. What effect does increased opportunity for oral and written verbalization in algebra have upon students' understanding?

As research findings become more widely communicated to teachers and other practitioners, the implications of research will have an increasing impact on classroom instruction. Some of the early research on students' misconceptions in algebra has already alerted teachers and textbook authors to ways of sharpening their presentations by including a broader range of examples, as well as nonexamples, in the discussion of key concepts. However, the means by which research results can be translated into effective classroom teaching are not always obvious:

- How can *knowledge of cognitive obstacles* that are encountered in the learning of algebra be used to improve classroom instruction?

In addition to general teaching strategies, a number of specific topic areas were identified at the conference as particularly important in algebra and greatly in need of continued research to inform our teaching:

- We need more explicit *models of instruction* in algebra.

    a. When and how should symbolism be introduced?

    b. How should we teach about variables and functions?

    c. What are the best ways to teach algorithmic skills?

    d. What are the best ways to teach thinking strategies (generalization, reversibility)?

    e. Which algebra topics are best taught through direct instruction? through problem solving?

Given the myriad ways that computers can be used in teaching algebra, from problem-solving tool to tutor, the need for investigating the range of possibilities and developing prototypical materials was reiterated a number of times by conference participants:

- We need models of methods for *integrating computers* into the algebra classroom.

Finally, an established source of research inspiration—the expert teacher—was suggested as a pertinent and potentially valuable resource in research on teaching algebra:

- What and how can we learn from *expert algebra teachers*?

    a. How do we identify them?

    b. How do we determine what it is that they do, and for whom?

Algebraic Thinking

Closely linked to issues of learning and teaching in algebra are questions related to thinking and reasoning. One point that has interested researchers for some time is whether thinking processes in algebra are essentially the same as those used throughout mathematics—and in other contexts, too—or whether certain thought processes are fundamentally different and unique to algebra:

- What *dimensions* of algebraic thinking can we identify (e.g., knowledge of structures, use of variables, understanding of functions, symbol facility/flexibility, generalizing, inverting and reversing operations and relations, ability to formalize arithmetic patterns, etc.)?

  a. What kinds of thought processes are involved in various algebraic topics?

  b. What kinds of thinking processes are required to apply algebra to problem situations?

  c. What are the effects of studying specific topics on students' facility in algebraic thinking?

Once the various dimensions of algebraic thinking have been identified, the next question is how to facilitate the development of these processes:

- How does/can a given dimension of algebraic thinking *develop*?

  a. What skills/concepts mediate algebraic thinking?

  b. What instructional strategies promote the development of certain dimensions of algebraic thinking?

  c. Are there particular types of (word) problems that stimulate the development of algebraic reasoning?

Affect

Issues related to affect were not addressed at the conference as explicitly as they probably should have been. Affective considerations arose throughout the discussions but usually in tandem with cognitive considerations, as in "the cognitive and affective effects of (whatever)." Given the fact that algebra is the "critical filter" that prevents many students from pursuing mathematics beyond arithmetic, the issues related to affect in algebra clearly merit separate attention. As just two examples of the kinds of questions that need to be addressed, the following were suggested:

- What are the affective effects of *different pedagogical approaches* to algebra?

- What are the affective effects of *different representation systems* in algebra (e.g., natural language, graphing, tabular data, alphanumeric symbolism)?

The prospect of a significantly enhanced role for technology in the teaching of algebra raises two other questions that relate at least indirectly to affect:

- Does increased use of technology in the teaching of algebra have a *differential effect* on boys and girls with regard to achievement or attitudes toward mathematics?
- How does computer-assisted instruction affect the *sociology* of the classroom?

Representation

One of the major issues related to representation is the role of natural language in mediating the formation of cognitive representations of algebraic concepts. In research that treats algebra as a language, one of the basic questions is the relationship between thought and language:

- How does a particular representation system (e.g., natural language, coordinate graph, alphanumeric symbolism) influence *understanding* in algebra? Conversely, how much concept development needs to precede the *introduction* of various kinds of symbolism?

As indicated in the section on major themes, the process of translating verbal problems into algebra seems to involve more than the translation of one language into another:

- What is involved in the process of *translating* real-world problems into algebraic formalism? What is involved in the process of translating between symbol systems? How can we best assist students in making these translations?

With the advent of computer capability for providing feedback to students using linked representations, questions about multiple representations arise that parallel earlier concerns about multiple embodiments in the elementary grades:

- How do students learn to use and coordinate *multiple representations* of key concepts and procedures, such as unknown, variable, equivalence, relation, inequality, solving equations, and modeling quantitative situations?

  a. Is it better to have students work with multiple representations from the very beginning of a topic, or should they understand one representation well before another one is introduced?

  b. Does the division of attention between representations have a deleterious effect either on understanding links between representations or on performing algorithms?

Technology

Questions regarding the potential impact of technology on the learning and teaching of algebra dominated much of the discussion throughout the conference. Concerns related to technology appear in virtually every section of the agenda. The questions listed in this section are those in which the focus is primarily on the technology itself.

The controversy over the role of computer programming in the algebra curriculum provoked some of the most spirited discussion at the conference. Some participants gave persuasive arguments for expecting all students to do some programming (i.e., simple programming as a basic skill), while others stood firm in their conviction that, even for algebra students, it is enough to use computers as tools (as in driving a car). The only consensus reached was that there are important research issues on both sides of the debate:

- What effects does *programming* experience have on students' learning and understanding of algebra?
  a. What is the role of procedural/algorithmic representations in the development of algebraic thinking?
  b. Are some programming languages better than others for developing algebraic thinking?
  c. Is the developmental level of the student a significant factor in working with programming languages, as it is with natural languages?

The variety of ways in which computers can be used as tools is well illustrated in the curriculum/software development projects currently underway. We need research to assess the impact of these various modes of usage on students and teachers:

- How can the *effectiveness* of computers-as-tools be maximized through such means as:
  —using real-world algebra problems,
  —teaching old concepts (e.g., interpolation) in new ways,
  —developing different algorithms,
  —incorporating nontraditional topics like curve fitting or spreadsheets, and
  —working with muMath?
- How does the *role of the teacher* change with different modes of computer usage?
- How can computers-as-tools be used to develop *algebraic thinking* skills?

Related questions surround the use of intelligent tutors. Though these tools are quite different from problem-solving tools, they do offer advantages in certain situations:

- What are the best ways to use *intelligent tutors*—in instruction? in research?

Apart from issues related to multiple representations (see previous section), computers introduce issues related to the linking of multiple representations:

- What are the effects of working with *dynamically linked*, multiple representations?

  a. How are the learning and understanding of concepts and procedures affected by experience with multiple, linked representations?

  b. To what extent does experience with multiple, linked representations enhance students' ability to select and shift among appropriate representations?

  c. Does work with multiple, linked representations enhance students' ability to make appropriate generalizations and solve problems?

Some other issues that arose as conference participants reacted to a draft of this agenda are indicative of the wide range of research issues that technological developments will continue to inspire:

- What are the *implications of the shift* from the traditional, highly symbolic, logico-deductive approach to algebra to a more numerical, empirico-inductive approach made possible by the advent of computers?

- Does heavy reliance on technology *interfere* with the learning of some concepts?

- What are the effects of *procedure-capturing software* on students' ability to generalize and solve problems?

- What are the cognitive consequences of student participation in the construction and use of *notational* systems?

- What are the implications of *constantly available technology* (calculators) in contrast to intermittently available technology (computers)?

Curriculum Development

In keeping with the goal of improving instruction in algebra, one morning of the conference was devoted to the presentation and discussion of issues related to the curriculum. Some of the points raised were the following:

- What *algebraic topics* (both traditional and nontraditional) are appropriate for different populations of students? at what points in their school careers?
    a. How should these algebraic topics be sequenced?
    b. What algebraic ideas are appropriate in the elementary grades?
    c. What arithmetic/geometric experiences can be provided at various levels to anticipate the later study of algebra?

Computer experience with children as young as five years of age has shown that certain algebraic concepts (e.g., the notion of variable) are accessible to students much earlier than has generally been assumed. Such findings call into question our methods of teaching, as well as the scope and sequence of introducing algebraic ideas into the curriculum:

- What kinds of *learning experiences* (e.g., skill practice, computer programming, proof) should be incorporated throughout the pre-algebra and algebra curriculum?

With all of the flexibility that technology provides, it behooves us to be systematic in our investigation of overall strategies and specific tactics for curriculum revision:

- We need *models* for integrating text materials and computer software into instruction, testing, and planning in algebra.
- What are the effects on achievement and affect in algebra of such *curriculum features* as continual review, worked examples, increased emphasis on problem solving, and decreased emphasis on skill practice?

Testing

Recognizing the impact that testing has on curriculum and instruction, conference participants identified a number of concerns related to testing. Only the first question listed below is content-specific to algebra, but the other items merit attention because the algebra curriculum is at such a critical juncture in its evolution:

- How can *knowledge* of algebra be assessed, including such components as:
    —evaluating algebraic expressions and functions,
    —writing equivalent expressions,
    —relating one representation to another,
    —developing mathematical models,
    —acquiring function sense (parallel to number sense),
    —solving equations, systems of equations, and inequalities by various techniques,

—creating and/or carrying out algorithms, and

—reading and verbalizing mathematics?

- How can *affective factors* be assessed, including such dimensions as:

—persistence,

—motivation,

—independence, and

—appreciation of, and beliefs about, the nature of mathematics?

- How should tests *change* with the curriculum?

  a. How can we evaluate the effectiveness of new methods of instruction?

  b. What methods can be used for evaluating algebraic thinking?

  c. How can tests be developed to better inform the teacher and not just assess the students?

Teacher Education

At the closing session, the importance of issues related to teacher education was emphasized by several participants. As in the area of curriculum development, many of the concerns expressed are not restricted to algebra but are nevertheless highly pertinent:

- What are the most effective ways to *prepare* teachers to teach algebra?

  a. What is the optimum balance between content and pedagogy in teacher preparation?

  b. What is the best content preparation for teachers of algebra? Will future teachers of algebra need to know topics that are not currently in their basic programs, such as catastrophe theory, fractals, and the like?

  c. What types of pedagogical knowledge do teachers of algebra need?

- How can we keep *in-service* teachers up-to-date? How much and what kind of teacher training is necessary for teachers to internalize new ideas and new perspectives?

- How can we enhance the *professional development* of teachers?

  a. How can we get teachers more involved in curriculum development?

  b. How can we better communicate research implications to teachers?

  c. How can we encourage the teacher researcher?

CONCLUSION

There are many ideas presented in the agenda, some dealing with larger questions than others. At the conference, it was not possible to come to closure on the ideas that were discussed. They need time to be assimilated. Nevertheless, we hope that some of the points suggested may help individuals or small groups of researchers come to closure on their own research agendas.

The remaining papers in Part II of the monograph present critiques of the agenda from a mathematics education perspective, a cognitive science perspective, and a curriculum development perspective. Part III features three papers that discuss theoretical research issues that need to be addressed as we continue our pursuit of insight into the learning and teaching of algebra.

The Research Agenda in Algebra: A Mathematics Education Perspective

Lesley R. Booth

Department of Pedagogics and Scientific Studies in Education
James Cook University of North Queensland

Developing a research agenda in any area is no small task. Where the subject of algebra is concerned, this task takes on a particularly challenging aspect, for several reasons. Perhaps the most compelling of these reasons is the widespread and continuing difficulty with which students (and teachers) appear to confront the subject, and this despite two decades of curriculum reform and development. Why have these sometimes excellent curricular initiatives remained largely unsuccessful in improving students' understanding and appreciation of algebra? The question is not an idle one, since the reasons for the lack of success in these endeavours may well be critical to the potential value, in terms of successful outcomes, of any research agenda that is now proposed.

A second reason for challenge in this subject, and one doubtless related to the first, is the nature of the subject itself. What is algebra, and what are the essential features of algebraic activity that must comprise the goals we have for children's learning in this field? It is clear that views on this are changing, though one of the problems with the teaching of algebra has been that there was often a very unclear picture of precisely what the goals of learning in algebra might be. Failing a clear and convincing analysis in this regard, teachers and textbooks fell back on the surface features of algebra and contented themselves with teaching manipulative skills and the routine application of a few standard algorithmic procedures. The advent of technology has forced a reexamination of this approach. Indeed, if the introduction of computers into mathematics education does no more than make us rethink what we are about, this would be value enough!

A further reason that developing a research agenda in algebra is a challenge concerns the diverse and as-yet-uncoordinated wealth of research findings that already exists in this area. This wealth is by no means extensive in quantity of research but, rather, in quality and depth. It is somewhat surprising that, despite the particular difficulty with which this subject area has always been regarded, and despite the very central role of algebraic activity within mathematical thinking, there has been relatively little research that has focused specifically upon this area. This situation is of course changing now, as more educators are voicing concern over the way we teach algebra and are thus giving it greater priority—as witnessed by the choice of algebra as a priority area by the National Council of Teachers of Mathematics for the holding of a Research Agenda Project conference.

However, there is some research evidence already existing in algebra and in related areas—research which can greatly inform our understanding of children's difficulties in this subject and provide guidelines for rectifying these difficulties. Unfortunately, the essential aspects of this research evidence remain largely unreflected upon. In the desire to plan ahead and produce agendas for "new" research in this area, there is a real danger that much of value that is already in the research literature will remain essentially unexamined and, worse still, that valuable time and resources will be spent on reproducing the insights that are already there for the careful reader to construct.

A similar comment must apply to theoretical issues related to the learning and teaching of algebra. There is already much written about children's learning in general, and the learning of mathematics in particular, which can provide us with extremely useful frameworks to guide our thinking, and hence our research, in algebra. We do not really need new theories; what we *do* need, in order to develop guidelines that will assist our investigations, is a far more critical and in-depth analysis of the theoretical perspectives that are already there. At the same time, we need to plan our research so that it can also inform the theoretical perspectives within which we work.

THE ROLE OF THE RESEARCH AGENDA

What does all of this have to say for our proposed agenda? In the first place, it means that the agenda faces several extremely important tasks: (a) It must present a framework which will permit a careful examination of all the factors that underlie students' (and teachers') difficulties in algebra and which will indicate possible teaching solutions; (b) it must also look to the reasons why earlier curriculum initiatives have not been successful, in order to indicate guidelines for implementing the findings from the research suggested in the agenda; (c) it must take on the responsibility of defining the school algebra of the future and, hence, what our goals for children's education in this subject might be; and (d) it must present an integrated picture of what research and theory have already told us, so that we have not only a baseline from which to work, but also guidelines to indicate potentially critical avenues for further investigation.

Clearly, this is an enormous endeavour and one that could not possibly be achieved in the space of four days, no matter how expert and dedicated the group. Appropriately, the conference chose to focus on the first and third of the tasks outlined above, paying explicit but necessarily brief attention to point (b)—issues related to the implementation of research findings—and to point (d)—overviews of existing research and analysis of the potential usefulness of some newer theoretical perspectives arising from cognitive science.

I say "appropriately" for several reasons. In the first place, research in mathematics education, unlike research in some other disciplines, must look to practical and useful outcomes. Although the insight into children's knowledge and learning of algebra is an outcome valuable in its own right, it is by no means enough. The critical feature of education research in general, and mathematics education research in particular, is that it should have a useful outcome, by which is meant that research in mathematics education *should result in an improvement in children's learning of, understanding in, perceptions of, and attitudes toward mathematics*. Furthermore, the results of such research must be seen to have the potential for relatively immediate applicability to the classroom for at least three reasons: Not only are there moral and economic obligations, not only do the outcomes from educational research normally require implementation within the context in which they were established, but also the status of mathematics education research requires applicability.

Important though the first two reasons are, the third reason should not be overlooked. Research in mathematics education is still a relatively new field. It has not as yet established its reputation nor earned any kind of public or political (in the general sense) recognition. Until it does so, it will have relatively little power to influence questions of curriculum, planning, or policy in mathematics education—or, of course, to attract significant funding. What will help it to achieve the level of recognition required is the demonstration that research can produce outcomes that have a positive effect in the classroom. In order for mathematics education research to have an effect in influencing policy, this demonstration cannot be delayed too long.

This goal of achieving demonstrable and positive improvements in children's mathematical understanding and performance, and within a relatively foreseeable timespan, has immediate consequences for the *kind* of research we do and for the *ways* in which we do it. In the first case, we must select research questions that can lead to practical consequences and which teachers and others most influential to the child's learning perceive as being important; in the second case, we must choose methodologies that are convincing to teachers and which also might involve them. The attention paid to point (a)—the development of a framework of practical questions to guide research in the learning and teaching of algebra—shows that the developers of the agenda recognise the fundamental and important issue of applicability in mathematics education research.

The second appropriate direction of focus is point (c), concerning the nature of algebra. Indeed, this was an issue that permeated the entire conference, and rightly so. It cannot be stated too strongly that the most important aspect in defining any research agenda must be the delineation of the goals towards which that research looks. This is critical in any research area; where the learning of algebra is concerned, it takes on a particular importance for the reasons already outlined. Views of the nature of algebra are changing. The introduction into the classroom of increasingly more powerful calculators and computers is dramatically changing what students can do in this subject. Likewise, our expectations (and those of society) about what students should be able to do and understand are also changing. All of these factors necessitate an urgent reappraisal of what the goals of learning and teaching algebra should be. This reappraisal must precede the initiation of further research in this area; failure to do so will result in enormous (and costly) research efforts being directed toward the better teaching of curriculum material that is no longer relevant in the classrooms and society of the future. It is this concern that underlines the attention paid in the agenda to the question of the nature of algebraic learning in the future.

So how has the research agenda in algebra met these two primary challenges? To what extent are the research questions proposed relevant to the goal of improving children's understanding and experience in algebra, and what guidance does the agenda provide concerning the future nature of the algebra curriculum?

IMPROVING THE LEARNING AND TEACHING OF ALGEBRA

To improve the learning and teaching of algebra, we must first ask what research questions might be of central importance to the achievement of our goals and also of particular concern to teachers. In fact, the questions that most concern teachers are likely to be those related to their *goals* (What should we be teaching?), their *starting points* (What do children know? What difficulties are they likely to experience? How do these concepts develop? Are the children at an appropriate stage of development and experience?), and the *means by which they can best achieve the goals* (What are good ways of helping children learn this?).

If we examine the research agenda itself more closely, we will notice that the same three concerns—goals, starting points, and methods—underlie the areas for research delineated in the agenda. It would take only a cosmetic rearrangement of topics in order to make this point more clearly. An examination of the agenda also shows us how closely the proposed research questions meet the requirements of delineating areas that have the potential for practically useful outcomes and thus are likely to interest teachers. The agenda can, of course, be equally useful in helping us to identify priorities. For example, a careful examination quickly reveals the relatively greater number of "how" questions, relating to the methods aspect of our framework (e.g., how can we best achieve our goals, given the identified starting points). This undoubtedly reflects our need, as a community, to achieve useful results in mathematics education research. It may also reflect our recognition of the importance of questions that will be of particular concern to teachers, namely questions of methodology in teaching. Perhaps, too, it shows us that we recognise the need to find some immediate answers: We are not willing to put off the "how" questions until we have established answers to "what goals" and "what starting points."

However, an over-concern with finding methods of solution can be counterproductive if due regard is not also given to the goals of teaching algebra and to the bases, in terms of child and teacher needs and abilities, from which those methods must derive. Overlooking the goals is likely to result in the direction of research initiatives toward the better teaching of outmoded curriculum material, as we have already noted. Overlooking needs and abilities may result in failure of the proposed solutions simply because the required foundations have not been laid and, furthermore, may result in the long-term discreditation of approaches that might in themselves be extremely valuable. Consequently, the emphasis manifested in the agenda on questions of how to improve the teaching of algebra may be detrimental to long-term success in this area, if research proceeds without careful regard for the goals and starting points of teaching. In fact, research is unlikely to proceed in such a manner, and herein may lie the source of potential problems. In the absence of externally suggested answers to questions of aims and starting points, researchers will assume their own. Indeed, we can see clear indications of implicit assumptions in many of the agenda items.

Should research on methods await the results of research and conceptual analysis in the area of goals and foundations? Regardless of what the most academically acceptable answer might be, the requirement for immediacy in applicability discussed earlier demands that at least some research look now toward the issue of finding more successful teaching procedures in algebra. In pursuing such research, however, it is essential that the underlying assumptions, in terms of goals and starting points, be made fully explicit and, wherever feasible, provision for alternative possibilities be built into the research procedure. For example, a study that sets out to examine the effectiveness of a particular teaching approach might take care to incorporate a careful analysis and documentation of the antecedent conditions and indicate alternative pedagogic outcomes that might ensue. By these means, both the terms of reference and, hence, the potential of the research can be made accessible to immediate public scrutiny. In addition, the research itself can be informed by the data derived from the study of the alternatives it incorporates.

Considerable attention has been paid in the above discussion to the importance of teacher concerns as a criterion for assessing the research questions proposed in the agenda. There are at least two very important reasons for this. First, teachers are very well placed to have insight into the questions likely to be crucial to the goals of improving children's experiences in algebra; second, under current educational practice, it is the teacher who selects and designs the child's learning experiences, and who is therefore responsible for interpreting the curriculum, the requirements of society, and the examination system on the one hand, and the needs and abilities of children on the other hand. Consequently, in order that its findings and recommendations may achieve successful implementation in the classroom, it is necessary that mathematics education research be planned to work through the teacher as the main agent of change.

Since the achievement of the desired goals of mathematics education research requires the participation and cooperation of the teacher, researchers must consider how this participation and cooperation can best be ensured. There are at least two aspects that are of importance here, both of which were mentioned earlier. In the first place, research must seek to answer those questions of most central concern to teachers and provide the means to create and implement solutions; in the second place, attention must be given to the kinds of research methodology adopted. This second aspect is in part related to the first. For example, the requirement that research provide teachers with the means of solving instructional problems will often carry with it the need for teachers to obtain first-hand understanding of children's difficulties, misconceptions, ways of learning, and interactions with different instructional procedures. The requirement is thus for teachers themselves to be involved in the research process, or at least for the methodology of the research to be accessible to them. A further requirement, of course, is for the research methodology to be acceptable to teachers in terms of its perceived validity and range of applicability. Teachers who are unconvinced by the methodology adopted in a piece of research are not likely to be persuaded by the findings it produces.

This issue of methodology and the involvement of teachers and other decision-makers in the research process has been only indirectly addressed by the agenda. Nevertheless, these aspects remain critical to the eventual success of the proposed research programme and therefore require serious and urgent consideration by all those engaged in mathematics education research.

THE NATURE OF ALGEBRA

Perhaps the most fundamentally important question to which the agenda has addressed itself is the question of the nature of algebra and our goals in teaching it. A reevaluation in this regard has, of course, been stimulated by the rapid and significant technological developments that have opened up totally new possibilities in what students can learn and have also brought into question the relevance of "by hand" manipulations and simplifications that machines can handle far more efficiently. However, the move towards reappraisal of the algebra curriculum has also developed quite independently of technology, as an outcome of research into children's understandings and misunderstandings in this topic. Without the technology, it would be far more difficult to implement the recommendations from research, both from the technical point of view and, more importantly, from the viewpoint of providing a convincing rationale for reducing the emphasis on manipulative skills. The use of calculators and computers gives us not only the means for doing this but also the justification, in a form that is more convincing to teachers, parents, and employers alike. Indeed, the introduction of this technology has created a demand for curricular initiatives that take full advantage of its potential; the possibilities for successfully implementing curriculum reform in algebra are therefore greatly enhanced. In order to implement reforms, though, we need information from research concerning children's learning in algebra. Without such information, we lack valuable guidelines as to how we can make best use of the technology we have.

The essential dimension underlying current views on the nature of the algebra curriculum is the move from an emphasis on manipulative skills to an emphasis on conceptual understanding and problem solving, that is, a move from doing algebra to using algebra. This shift necessitates that students develop early on an appreciation of the meaningfulness and purpose of algebra and its use as a thinking tool. The role of technology in developing this appreciation is not only to provide a means for handling the required manipulations—thereby leaving the student free to concentrate on the problem-solving aspects—but also to provide a powerful aid to concept and strategy development in algebraic thinking. As a consequence, the structure of the algebra curriculum is reversed, with attention to problem solving preceding examination of the ways to handle the solution. Attention thus needs to be given not only to *what* is being represented, in terms of the underlying structures and relationships in problems (the semantic aspects of algebra), but also to *how* these are represented (the syntactic aspects). An important feature of algebraic thinking is the development of flexibility with regard to mode of representation; students therefore need to be able to recognise different forms of representation, know what advantages each has to offer, and be able to translate freely among them. Here, *different modes of representation* refers not only to different algebraic forms of expression, but also to graphical, tabular, and other means of representation.

In addition, attention must be directed towards the development of algebraic processes, both at the level of general heuristics and at the level of particular procedures. In line with the change in emphasis of the curriculum, will be a reassessment of the kinds of procedures stressed. Both the involvement of technology, and the research evidence as to the kinds of procedures children use naturally, point to a greater use of recursive techniques and a possible abandoning of some of the deductive (and restricted) procedures that form a large part of the current syllabus. Further curriculum features to which the agenda points include a focus on functions as the cornerstone of algebraic activity and the incorporation of multivariate cases in which the relationships are not restricted to linear ones.

Specific details of these proposed changes in emphasis in the algebra curriculum are given in individual conference papers. The importance of the implications of these for the agenda as a whole cannot be too strongly stressed. These proposed changes imply an urgent need to explore the limits of what technology can do with respect to algebraic thinking and to investigate the whole question of the role of multiple representations and their interactions with concept formation and thinking styles. Moreover, these changes also redirect the investigations we pursue into children's learning and thinking in algebra. As such, these new goals must influence all of the research upon which we now embark.

SUMMARY

The above analysis has served to focus our attention on certain principles that might guide our general approach to a research agenda. First, the central goal in identifying priorities and suggesting guidelines for research in mathematics education is the selection of research questions that will help improve children's understanding and experience in mathematics and that offer the prospect of fairly immediate applicability. The areas for research delineated by the agenda make a substantial contribution to the fulfillment of this goal. Second, urgent attention must also be directed to the nature of algebra and hence to the goals that we have for children's learning in this subject. An important beginning in this direction has been stimulated by the agenda.

Though the agenda has concentrated primarily on the above two matters, we must not lose sight of the remaining aspects critical to the planning of research in algebra. Researchers must take responsibility for considering ways in which the outcomes from their research might achieve impact in the classroom. This is necessary not only because the aims of educational research demand it, but also because the professional status of research in mathematics education requires it. Successful implementation of research findings relies upon the cooperative involvement of teachers. This means that researchers must select those research questions that are important to teachers and that have the potential for fairly immediate application; researchers must also consider ways of more directly involving teachers in the research process, with possible consequences for the kind of methodology adopted.

In addition, researchers must examine the context in which their research proceeds and delineate the assumptions made with regard to the goals of algebra instruction. Researchers must also pay full attention to the starting points of their research, including both the relevent research literature and the appropriate theoretical perspectives. Research cannot progress effectively unless it is based upon the insights already available. Furthermore, researchers must accept the responsibility for not only making explicit the theoretical orientations guiding their research but also planning their research to permit reflection upon the theoretical underpinnings. Such research-based modifications and refinements of theory might then permit the elaboration of more powerful theoretical frameworks, with eventual benefit to the design of more powerful research and the provision of more effective approaches to the classroom teaching of algebra.

The Research Agenda in Algebra:
A Cognitive Science Perspective

Matthew W. Lewis
Department of Psychology
Carnegie Mellon University

My charge was to write a reaction to the research agenda in algebra as viewed by a member of the cognitive psychology and artificial intelligence (AI) research communities. I have two additional goals:

- To convey some of the excitement over the ideas and the convergence of different research perspectives and paradigms that occurred during the development of this agenda, and

- To exhort any and all others interested in how algebra is learned and taught to think hard about how this agenda can be implemented in all areas of research and practice—in the classroom, in teacher education programs, in psychology departments, in schools of education, in curriculum development offices, in areas of computer science research, and in other contexts.

I will briefly address these two goals first. Then I will present a brief overview of the main interests of cognitive psychology and AI, as a backdrop for my reaction to the agenda.

THE DYNAMICS OF GENERATING AN AGENDA

The problem is the following:

> How do you take 28-plus highly motivated, highly opinionated, and highly energized people from widely varying research traditions, paradigms, and countries, put them in conference rooms for 3.5 days (and nights), and come out with a cogent, organized set of research issues that represent our best ideas for advancing our understanding of algebra learning and instruction?

This was the ill-defined task that faced the conference organizers. At the end of the first day I was a bit dazed at the spectrum of issues raised; I saw little hope for any convergence of concerns and ideas. At the end of the third day a convergence had nevertheless emerged: What finally characterized the meeting was a strong acceptance of the pluralism of research perspectives and concerns regarding algebra education. On paper, the agenda reads fairly straightforwardly, without testimony to the dynamics that went into the arguing and shaping of the topics. However, the agenda should *not* be viewed as a set of well-defined goals set in stone. It is anything *but* that. Instead, it should be viewed as a dynamic set of shared concerns that the conference attendees thought would help us better understand our students and their learning processes. It is meant as a collection of stimuli for further ideas, as indicated in the following section.

GETTING A PIECE OF THE ACTION

As we all know, knowledge and problem solving in algebra are huge and complex areas. As we also know, algebra is a primary "gatekeeper" to the later study of technical fields. Research work in the area of algebra has been steady, but of limited volume and on a limited number of topics. To make some serious progress on the issues surrounding algebra, we need to make a concerted effort on many fronts. Here are some possibilities:

- *Algebra instruction in the classroom*: Consider collecting and documenting systematic errors or interesting mental bugs that seem to co-occur in certain students. A set of accounts of when students "see the light" for different algebraic concepts is needed and would be fascinating. Try getting other algebra teachers interested in forming groups for short- or long-term projects. For example, design and compare different approaches to teaching different concepts using different sets of examples. What kinds of questions did each approach elicit? Which groups were best able to handle new problems? Which approach fostered better retention? Experiment with different curricula and share the results with fellow teachers and the rest of the mathematics education community by publishing in journals like *Mathematics Teacher*. Present your findings to other teachers in your district. Present your findings at the annual NCTM meeting or at other professional meetings. The insights into algebraic knowledge gained from daily interaction with instruction are priceless and should be shared with the rest of us.

- *Teacher education programs*: Consider involving prospective teachers in the research aspects of algebra instruction, including running small, even informal, studies during their student teaching and other fieldwork. Ask which of these agenda items would be most helpful in keeping a new teacher thinking about innovation in instruction. Form discussion groups to push the perspective that teachers can also be researchers and that information from the classroom is critical to others in the field.

- *Schools of education and psychology departments*: Consider offering classes, both undergraduate and graduate, with the agenda as a focus. Projects for such a class could include developing related ideas and writing grant proposals for research. Seminars could allow students to carve off a piece of the agenda, do related readings, and lead focused discussions of an area in depth. Raise questions of testing and measurement with people interested in psychometrics. What is "understanding" and how might we get a handle on it? Encourage students to do thesis work in these areas. Then, encourage students and faculty to attend meetings and publish the results of any research.

- *Curriculum development offices*: Encourage and support the conduct of research in your schools. Analyze different texts and run comparative studies. Organize discussions with teachers and supervisors about the agenda and the questions it raises for curriculum planners. Communicate the results of your work.

- *Computer science research*: Consider how computers as tools can and will affect classroom instruction. How might graduate student theses address these issues? How can we develop and evaluate new computer-based tools to help teachers in the classroom and researchers in the laboratory? How might we best communicate our results to the rest of the community?

In order to make headway, we need to light and tend the fires of interest in a wide range of research topics, approaches, and contexts. Then, when interesting results are found, we need to communicate them to others who are interested. The prospect of a concerted effort makes the future look promising for making substantial inroads into our understanding of algebra learning and instruction.

WHAT ARE COGNITIVE PSYCHOLOGY AND ARTIFICIAL INTELLIGENCE?

Generally speaking, the goal of cognitive psychology is to formulate and test theories that model or explain how people perform cognitive functions—pattern recognition, attention, memory, problem solving, visual imagery, and language. However, cognitive psychology is not monolithic in approach; within the field there are many paradigms and perspectives.

Much of what is traditionally thought of as cognitive psychology is laboratory-based research. The experimental approach is based on trying to capture cognitive phenomena in carefully controlled situations so that different variables can be systematically varied, and theories can thus be tested. However, the more carefully controlled the experiment, the further removed from the everyday world is the cognitive phenomenon of interest. There is a trade-off in trying to balance the "richness and messiness" of the real world with the "impoverishment and tractability" of laboratory studies. Both settings have their costs and benefits, and both should be pursued.

Artificial intelligence has already been addressed in Thompson's paper (this volume), so I will not go into detail here. I will only highlight the difference between AI and cognitive simulation. If we want to simulate "intelligent" problem solving, an AI approach would be to write a computer program that used any techniques available to get the problem solved. In the case of cognitive simulations, the goal is instead to simulate the particular steps *that humans take* and the representations of knowlege that humans allegedly have in their heads as they solve problems.

Taking a cognitive psychology approach to the study of human thinking is neither "the best" nor the only alternative. Different approaches and perspectives give different answers, all of which contribute to our understanding of complex thinking processes.

REACTION TO THE AGENDA

From the perspective of a cognitive psychologist, this agenda is a gold mine of research questions. Several of the areas of the agenda address important questions at the heart of current debates: representation, learning, stages of development, transfer and interference of knowledge, and analogical reasoning. In addition, working on research questions in the domain of algebra has some distinct advantages. Problem solving in traditional high school algebra is "well-defined," that is, the problem to be solved is usually clearly specified at the outset, as are the form of the solution and the permissible transformations. This makes the study of problem-solving processes in algebra easier than in other fields because solution paths can be more readily followed and analyzed. Contrast the prospect of researching how students learn to solve algebra problems with researching how people learn to write fugues: How do you know when a fugue is a "good" fugue? How does the student or researcher know when the problem has been "solved"? Algebra is easier to study because it is well-defined.

In sum, work on many of the agenda items would provide needed research results to areas of cognitive psychology; conversely, tools of cognitive psychology can be applied to many of the research areas. There appears to be a rich role for cognitive psychology in the study of algebra learning and instruction, with payoffs for the fields of both mathematics education and cognitive psychology itself.

Content

Central to the question of what the content of algebra should be is the issue of what we want students coming out of algebra to be able to understand and do. Of particular interest in the content area is the advent of powerful symbol-manipulating calculators. These calculators raise the same issues as four-function calculators in the earlier grades, except for one major difference: The calculations performed by the calculators in the earlier grades have traditionally been computational and algorithmic, whereas algebra often requires the use of heuristics or some search for correct problem-solving steps. How important is learning to apply heuristics for algebraic problem solving?

Content in relation to calculators also highlights the question of what are "basic" skills in algebra? What is the difference between basic skills and the strategic knowledge involved in problem posing and solving? Is it enough to be able to understand the problem and input the equation into a calculator for solution, or does the student have to know how to distribute? From a cognitive perspective come the following questions: What is the relationship between skills and understanding in equation solving? Can one be achieved without the other? What are the implications for theories of knowledge representation and skill acquisition?

Finally, why do we teach word problems? This question is aimed at a general question that should help drive our research on algebra content: What are the general concepts and problem-solving skills (not restricted to algebra) that we want to develop in our students? How are concepts and skills built longitudinally and across content topics?

Learning

What are the cognitive processes involved in learning algebra? One of the jobs of cognitive psychology is to attempt to build a theory of complex skill acquisition. But what is "skill"? Is it the ability to solve equations? Is it the ability to read a word problem and set up the equation? Is it the ability to reason abstractly about relations between numbers? Or, is it all of the above?

Part of the interest that cognitive psychology has in this area is in the stages and the transition mechanisms involved in skill acquisition. What theories can we apply to make specific and testable predictions? One theory that is fairly general and appears to make strong predictions is Anderson's ACT* theory (Anderson, 1982). In general, where do different theories break down, and why? What would better theories encompass and explain?

The issues of misconceptions and mental "bugs" also call for the application of theory. When students are faced with problem-solving tasks that require new or weak skills, how do they cope? One theory, called "Repair Theory" (Brown and VanLehn, 1980), addresses certain classes of errors. Work on common errors in algebra has also been done by Matz (1982). However, neither of these articles addresses the connections between the underlying mathematical concepts and facility with skills.

The perceptual issues involved in algebra are also fascinating from a cognitive perspective. Some have argued that much of expertise may be linked to being able to see patterns in problems and knowing what action to take based on the patterns identified. How do different learners differentially "see" an algebraic equation or graph? How do they differ in how they break it down into different pieces or chunks? (How do they "parse" these displays in different ways?)

Another question of interest is in defining and exploring experimentally what "insights" are: Given that people are differentially good at achieving insights, how might we "teach" the skill of developing insight into problems? Does knowledge from cognitive psychology about how people learn to categorize items by features apply to the task of insightful learning?

Teaching

Since one of the basic activities of cognitive psychology is building theories of learning and problem-solving processes, a natural question is: Given that we have some preliminary ideas about how skill is acquired, how can we use these ideas to motivate instructional strategies? If we understand parts of the skill acquisition process, then we should be able to design instruction that has the highest probability of being efficient and effective: We can play to the strengths of our students' learning processes and avoid the weaknesses.

A simple example of weakness is the limited capacity of short-term memory. When there is too much novel material during a teaching session (more than approximately 7 items or "chunks" of information being attended to at once), then things will be forgotten, errors will be made, and learning will be hindered. Another example is mentioned in the agenda: Some research has investigated the mixture of examples (from good examples through weak examples to counterexamples) and the numbers of each to present in order to maximize the learning of a new concept. More related work, and specifically work in the domain of algebra, remains to be done to translate these kinds of findings into successful instruction.

Cognitive psychologists will also resonate to the suggestion of studying the teaching strategies of expert teachers at work. To what cues from students are these teachers sensitive? How do these teachers select problems to present to students? How do these teachers think about the strengths and weaknesses of individual students? There is a growing tradition of studying experts in a domain to map out their cognitive processes for problem solving. Research like this can provide a data base on which to base a theory and/or computer simulation. Simulation work, in turn, can give indications of the rigor and thoroughness of the theory, as well as pave the way for future computer-based tutoring systems.

Thinking

Of central interest in this area is the concept of "algebraic reasoning." There is much room here for good definitions and empirical work to establish instances of algebraic reasoning, as separate from other reasoning, and then to investigate the conditions under which different styles of reasoning are used. Where does algebraic reasoning occur naturally? Can we cause it to occur in the laboratory?

Once definitions are established, then the issue of how to teach algebraic reasoning arises. Can it be taught to all students, or are there important developmental stages that are prerequisite? What might be "critical experiences" for students, and how might we study the effects of such experiences?

Affect

Affect and motivation may be the most critical areas of research, and the ones regarding which cognitive psychology has the least to say. We all know that people cannot be forced to learn. Regardless of the cleverness of the curriculum, the sophistication of the technology, or the charisma and caring of the teacher, students who do not care, or who are too scared to care, will not learn. Curriculum, technology, and human caring certainly might improve motivation and encourage positive attitudes, but they will not get the proverbial "horse to drink."

The field of affect and cognition is growing but is still quite young. There is good work in this area that tries to tie together affect, motivation, and cognition, but we have a long way to go. Given that algebra can be very heavily laden affectively, it would seem like a ripe area for research. Collect and analyze verbal protocols from algebra-averse students to see where affect has the most impact: Is it in the comprehension stages, the planning stages, the problem-solving stages, or throughout? What is the source of the fear or anxiety? Is anxiety due to previous failures at being able to apply algorithms? If so, could the introduction of calculators or other tools help to ameliorate the anxiety and hence free up cognitive resources for solving the non-computational parts of the problem solving? In sum, affect and algebra seems to be another area ripe for good research within cognitive paradigms.

Representation

As mentioned earlier, the issue of representations of skill and knowledge is at the core of much of the work in cognitive psychology and AI research. How do people represent their knowledge of a domain or skill in their heads? The theory is that there is some underlying structure to the knowledge, whether that structure is in the form of many "frames" (organized units of knowledge), rules (sets of conditions under which to take specified actions), networks of connections (dense nets of interconnected pieces of knowledge and skill), or mixtures of all three. Do we have pictorial representations of knowledge in our minds? Can we, do we, call up an image of a graph and reason about it?

Interesting questions from a cognitive psychology perspective abound. What mix and order of presentation of representations will lead to the best recall and subsequent use of knowledge about equivalence? What errors are most easily remediated by which representation, and why? What cognitive processes do different representations support or not support? What kinds of representations ease the burden on our limited capacity for attending to multiple aspects? Each of these questions has many possible theoretically based empirical studies behind it.

Technology

The development and use of technology to assist in algebra research and education offers both the possibility of tools for teachers, as well as tools to test theories of how people acquire complex skills. Using computer-based tutoring systems or guided-discovery environments, it is possible to carry out controlled studies of the acquisition of complex skills. The programs can provide guidance based on a set of theoretically specified principles. The instruction thus provided may vary from student to student, but the underlying principles will be manifested in the same way for all students. Although teachers are far better and more flexible tutors than are any computers, their very flexibility makes it difficult for them to provide controlled instruction to two experimental groups, if the goal is to test a theory. Carrying out experiments on computers not only provides better control but also allows the gathering of rich reaction-time and error data.

Technology also allows us to automate *adaptive testing*, a method of testing that varies the questions presented to each student, based on the student's knowledge and responses. This capability permits faster and finer grained assessments than are possible with standard pencil-and-paper tests. Ways of organizing the hierarchies of skills and concepts to be tested and methods of assessing attainment of these ideas are currently major thrusts of some educational testing firms.

There is growing interest among cognitive psychologists in the area of computer programming. What makes up programming expertise, and how can we study it? There are data that suggest that the best predictors of programming ability are mathematics achievement scores. How is programming ability different from other skills, mathematical or non-mathematical? What skills do and do not transfer to programming from other problem-solving domains—like algebra—and why?

Curriculum Development

Analysis of algebra texts shows that there are many skills that authors want to convey but that are still left implicit in the curriculum. Students must infer the rationale for certain steps in problems, and those inferred steps may go astray. Breaking skills down into explicit skill hierarchies through task analyses is one area of cognitive psychology that has become fairly well developed. Theories of skill acquisition should make some testable recommendations regarding the integration of skill presentation with computer-based skill practice.

Yet another area of research interest related to algebra is the work with imagery and visual displays that seems to explain why the use of graphs may be "worth 10,000 words" (Larkin & Simon, 1987). Work on the differences between what more- and less-able students do as they study, and later use, worked examples during problem solving seems to be of relevance to algebra learning. There is also the beginnings of a literature on word problems, both arithmetic and algebraic. What are the cognitive processes involved in understanding, setting up, and solving word problems, and what mix of problem contexts might be helpful? These questions relate to the point of how to design curriculum to maximize the analogical use of knowledge gained. Work is underway to analyze the processes of analogy and how to foster its use in problem solving, possibly through curriculum design.

Testing

Here again the issue is raised of what is "knowledge of algebra" and how can we assess it? As indicated above, cognitive psychology can offer tools for analyzing knowledge and designing instruments to measure it. We can help with answers to questions like these: How can data from written protocols be organized to facilitate interpretation of the assessment? Which factors tested will be the best predictors of performance in various areas, and why? Cognitive psychology has little to say about how to assess important but poorly defined concepts, such as beliefs about the nature of mathematics or an appreciation of the "art of mathematics."

Teacher Education

When all is said and done, education takes place when students are in classrooms with teachers on a day-to-day basis. Teacher education and in-service training are the direct avenues for affecting the way algebra is taught in the future. Part of my exhortation at the beginning of this paper was aimed at getting teachers and soon-to-be teachers directly involved in research of all kinds.

Cognitive psychology needs more input from teachers. Experiences during the course of my own research projects have reinforced the necessity of including classroom teachers in our research teams. The "reality therapy" of having a teacher's perspective is invaluable to any cognitive research that claims to have some application to what happens in classrooms. It is far too easy to be caught in the trap of "ivory tower" idealizing about what happens in real classrooms. Having access to teacher expertise tends to minimize that idealizing.

Since our current project involves developing computer-based tutoring systems for algebra, we have had the opportunity to invite groups of in-service teachers in for dialogues and demonstrations. This provides a productive cross-fertilization of ideas in both directions. We gain from the comments and insights of the teachers, and they gain a different perspective on their teaching area. The same exchange might be useful for students in teacher education programs.

SUMMARY

I began with three goals: (a) to attempt to represent the perspectives of my field in reaction to a research agenda, (b) to convey some of the excitement I felt over the ideas and the convergence of different research perspectives and paradigms that occurred during the development of this agenda, and (c) to exhort any and all others interested in how algebra is learned and taught to think hard about how this agenda can be implemented in all areas of research and practice and to find pieces of the puzzle/quilt to work on.

The agenda was the product of a very dynamic meeting. It was our hope that putting together an agenda could stimulate thought *and action*. There are a plethora of interesting questions regarding algebra to be asked and pursued. We rarely find definitive answers in education, but it is through the pursuit itself and the communication of efforts and ideas that optimal progress is made.

REFERENCES

Anderson, J. R. (1982). Acquisition of cognitive skill. *Psychological Review, 89*, 369-406.

Brown, J. S., & VanLehn, K. (1980). Repair theory: A generative theory of bugs in procedural skills. *Cognitive Science, 4*, 379-426.

Larkin, J. H., & Simon, H. A. (1987). Why a diagram is (sometimes) worth 10,000 words. *Cognitive Science, 11*, 65-99.

Matz, M. (1982). Towards a process model for high school algebra errors. In D. Sleeman & J. S. Brown (Eds.), *Intelligent tutoring systems* (pp. 25-50). New York: Academic Press.

ACKNOWLEDGMENT

This research was supported in part by the National Science Foundation under NSF Award No. MDR-84-70337. Any opinions, findings, conclusions, or recommendations expressed herein are those of the author and do not necessarily reflect the views of the National Science Foundation.

The Research Agenda in Algebra: A Curriculum Development Perspective

Sidney L. Rachlin
The Curriculum Research and Development Group
University of Hawaii

In a spirit of stimulating new directions and collaborations, the NCTM Research Agenda Project Conference on the Learning and Teaching of Algebra brought together cognitive psychologists, mathematicians, mathematics educators, researchers, evaluators, and curriculum developers in a true wedding of minds. As with any good wedding, the celebration featured something old, something new, something borrowed, and something to think about.

SOMETHING OLD

The success of collaboration in early number research provided the dream of what might eventually be accomplished in research on the teaching and learning of algebra. However, the initial discussions quickly revealed some significant variation among the participants' perceptions of algebra. For starters, it was difficult to agree on a definition of *algebra*. Yet, there was a strong consensus that there has been little change over the years in what has been taught as algebra in the high schools.

Since algebra was first made a college entrance requirement by Harvard in 1820, its position in the secondary school curriculum has solidified. During the period 1820-1900, the topics of algebra were literal equations and formulas, numerical equations and word problems, fundamental operations with rational expressions, powers and roots, and factoring polynomial expressions (Osborne & Crosswhite, 1970). By the end of the 19th century, the topic of factoring was at the peak of its role within the high school algebra curriculum. From 1900-1920, the topic of graphs entered the curriculum and found a foothold as a means of integrating geometry and algebra. The current growth in prominence of graphing and functions and the corresponding downplaying of factoring, powers, and roots are continuations of a gradual process that has actually been in motion for some time.

While the content of algebra has remained fairly stable over the last century, the suggested methods of teaching it have varied along several dimensions. Cajori's survey of mathematics teachers in 1890 (cited in Osborne & Crosswhite) revealed that teachers even then were opposed to what they saw as an overemphasis on manipulative skills and were calling for a meaningful treatment of algebra that would bring about more understanding. These calls have been echoed in waves of reform documents from that time until today—with proposals ranging from laboratory approaches that would encourage an inductive learning of algebra, to real-life applications that would demonstrate the relevance of algebra. Although most algebra texts of the first half of this century paid lip service to having students learn "why" as well as "how," the emphasis in the texts was on computational algebra. The common sequence for instruction identified by Osborne and Crosswhite was definition, illustration, rule and example, drill, review, and speed tests.

The first half of this century was marked by a tendency to separate what little research existed from decisions about the curriculum. This tendency continued throughout the "new math" reforms of the 1950s and 1960s and is still reflected today by the limited attention to curriculum formation in the proposed research agenda. Curriculum has generally been based on theoretical decisions made by knowledgeable adults regarding what students should learn, in what order, and for what purpose. In practice, research has had little, if anything, to do with this decision-making process. Curriculum research has usually been viewed in its most narrow sense, as the evaluation of a finished product. Changes made near the end of the development process tend to be superficial since the author's primary goals have presumably been met as part of the driving force in the production of the text.

We should not attribute to research those curriculum decisions that are basically philosophical, such as decisions to maximize the use of technology as a tool for learning or to include more structure, precision of terminology, or applications. On what scale can we compare curricula based on different philosophies? If one author considers factoring an unnecessary evil and holds off on it in the hope that it can eventually be discarded, but another author considers it a major component of the ninth-grade curriculum, do we compare the two curricula with items that measure students' ability to factor? Do either of these curricula have a research base that guides the selection of tasks or the organization of instruction within their unique philosophical perspectives?

SOMETHING NEW

The RAP wedding of the minds was also new in several ways, not the least of which was the polygamous nature of the marriage. True, cognitive psychologists and mathematicians did meet together during the last major wave of curriculum reform, but this time they were joined by a new area of specialization—mathematics education. For mathematics educators, research on the teaching and learning of mathematics is the lifeblood of their profession. Although psychologists had previously urged mathematicians to reflect some concern for the learner in their new curricula, mathematics educators brought to this conference a new concern for the teacher as learner.

Curriculum developers have for some time acknowledged the need for teacher training, but this training has tended to be as rote as the procedures it attempts to eradicate (Siemon, 1987). What was new at this conference was the realization that the teacher and, indeed, the teacher educator also learn through new curricula.

> A curriculum is more for teachers than it is for pupils. If it cannot change, move, perturb, inform teachers, it will have no effect on those whom they teach. (Bruner, 1960/1977, p. xv)

The content of curriculum change and the process of curriculum change represent distinct spheres of knowledge and expertise, both of which must be present and integrated in any curriculum reform effort. Regardless of what content society ascribes to algebra, there is a need for research on the learning and teaching of the curriculum at two levels—that of the students and that of the teachers.

Another new feature in today's marriage was the impending impact of technology on the existing curriculum. While everyone recognized the power of the new technology as an agent of change, the conference papers reflected confusion over the future use of technology in the learning of algebra. On the one hand, we were told of numerical approaches to applications that may well change the way that our students conceive of solving equations; on the other, we were told of efforts to design extremely powerful intelligent tutoring systems to model the rote equation-solving procedures found in today's texts.

According to Fullan (1987), one way of implementing change in the curriculum is through "brute sanity." What can be more natural or rational than advocating a change that you believe in and are in a position to introduce? The only danger in this, warns Fullan, is the tendency to "overlook the complexity and detailed processes and procedures in favor of [the] more obvious matters of stressing goals, the importance of the problem, and grand plan" (p. 5). As we rush forward to design a technologically advanced algebra curriculum, we must be careful to include research from both the learner's and teacher's perspectives to determine what aftershock may result from the new curriculum.

In a study of college students' understanding of intermediate algebra (Rachlin, 1982), I posed the following problem:

What real number added to $\sqrt{3}$ equals $\sqrt{6}$?

All of the students I interviewed initially wrote $\sqrt{6} - \sqrt{3}$, but most of them went on to further "simplify" their answers. One student pulled a calculator out of his book bag, keyed in $\sqrt{6}$, and wrote down 2.4494897. Then he keyed in $\sqrt{3}$ and wrote down 1.7320508. Finally he keyed in 2.4494897 − 1.7320508 to obtain 0.7174389 as his answer. When asked if this was indeed the real number that added to $\sqrt{3}$ equals $\sqrt{6}$, he replied, "Sure it is; I'll prove it." He picked up his calculator and punched

0.7174389 + $\sqrt{3}$. With a surge of confidence, he announced that the answer was 2.4494897, which he had originally shown was $\sqrt{6}$. Q.E.D.

Surely it is not the technology that is at fault here. Computers and calculators are, of course, just tools. It is not the new technology but, rather, the enlightened users of that technology who will design future algebra curricula. It is the responsibility of researchers to learn what additional understandings and misconceptions students develop as a result of using this technology.

One additional concern raised by this story relates to the place of the real numbers in algebra curricula of the future. As some have scorned the teaching of fractions in the age of the calculator, so too may we soon hear of the real numbers coming under close scrutiny. After all, applications of mathematics typically require approximations appropriate for a given situation. While it may seem ridiculous and unrealistic to talk of deleting the real numbers from the high school curriculum, remember that we easily taught our students to ignore the computer errors created by BASIC programs that list the squares of a series of numbers. The value of π may yet be set at 3.1 by a future state legislature!

SOMETHING BORROWED

Even a casual glance at the research agenda on algebra will show that much of it has been borrowed from other content domains. In fact, if we were to treat the word *algebra* as a variable, we could apply the agenda to many other topics of mathematics. But rather than determining which of the agenda items may be generalized, it may be more instructive to determine which agenda items are unique to algebra. The general nature of the algebra research agenda is a natural outcome of the marriage of minds. For mathematicians, mathematics educators, researchers, evaluators, and curriculum developers, algebra is but one of many areas of mathematics with which they interact. For cognitive psychologists, the domain of the variable *algebra* may include non-mathematical content.

It was not that long ago that publishers were seeking to "teacher-proof" their texts. Although all of the RAP wedding guests agreed that the role of teaching in the learning of algebra is generally not very well represented, not all of them were willing to buy the cause-and-effect relationship that emerged in the research agenda—that the ultimate goal "of all the agenda items is to improve our understanding of how students learn algebra and thereby to improve our classroom teaching of algebra." Some participants suggested that research on the learning of algebra should be interfaced with research on the teaching of mathematics. As one promising example of the melding of research on students and teachers, these participants borrowed from the current work in early number research the Fennema, Carpenter, and Peterson (1986) model for curriculum development (see Figure 1).

Models such as this help to clarify the notion that curriculum is far more than a task environment or a set of problems. Teachers play a large decision-making role in the implementation of curriculum. A better understanding of the role of students' cognitions and behaviors in their learning is a necessary but not sufficient condition to improve the learning of algebra. To implement change in the curriculum we must understand the nature of teachers' beliefs and cognitions and the roles these beliefs and cognitions play in the decisions teachers make as they present the new curriculum to their students.

> Change = Learning. Successful change, that is, successful implementation, is none other than learning, but it is the adults in the system who are learning along with or more so than the students. Thus, anything we know about how adults learn and under what conditions they are unlikely to learn is useful for designing and carrying out strategies for implementation. (Fullan, 1987, p. 7)

Any proposed change must be understood, accepted as necessary, and considered feasible by the teachers who will implement the change. Teachers and curriculum developers must be active learners in the curriculum implementation process. Algebra curriculum and instruction have generally been designed by knowledgeable and successful adults under the assumption that all students and all teachers think the same way. Further, these adults apparently assume that the way students and teachers think parallels their own thinking. In addition to research that helps us understand the diversity in students' thinking, we also need to understand the diverse range of prior knowledge, experience, beliefs, attitudes, and ways of processing information that teachers bring to the change process—if curriculum change is to occur (Siemon, 1987).

Still, there is something lacking in a model that uses research to provide input on the learner and the teacher separate from the impetus for the proposed curriculum change. The Fennema, Carpenter, and Peterson model

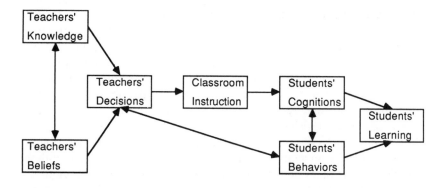

Figure 1. The Fennema, Carpenter, and Peterson model for curriculum development.

for curriculum development does not indicate their own role in the change process. These teacher educators/researchers collected and analyzed data on learners and teachers and designed an in-service program to change the teachers' knowledge and beliefs with regard to the learning of first-grade mathematics. In so doing, they were the curriculum developers, and it was through their teacher training that they sought to have their curriculum implemented. There is a lingering image of the knowledgeable adult gathering research data and then going off to create a new curriculum. Curriculum research is again treated as a posttest form of evaluation. What is lacking is a research paradigm that provides more dynamic bases for developing curricula and modifying instruction in both the student and teacher dimensions.

SOMETHING TO THINK ABOUT

From a curriculum development perspective, the most significant shortfall of the RAP algebra research agenda is its narrow view of what research on developing an algebra curriculum might entail. Figure 2 provides a foundation for a dynamic research model for curriculum development. The model should be viewed as a gestalt—a collage, which viewed in its separate parts loses much of its meaning. Included in the model are three triads: the curriculum developers, the teachers, and the students. Each triad consists of the individuals' beliefs, cognitions, and behaviors. Research in the process of curriculum change begins with some a priori knowledge of the cognitions and beliefs of each of the curriculum participants. It is through the behaviors of the curriculum developers, teachers, and students that we obtain a measure of the learning that is taking place.

The model is also like a fractal—in each triad there are complementary and conflicting forces for curriculum change. For example, algebra is only one of the students' academic courses. The students are also integrated into science, English, and social studies curriculum models. Teachers teach more than one subject and thus operate within a crystalline complex of intertwining triads. Curriculum researchers must be aware of the variety of uncontrolled forces affecting the environment at the same time that they focus on the interactions among the three triads: the classroom instruction, the direct interaction with students, and the teacher training (through in-service courses or teacher's guides). Beyond the separate, isolated studies within the teacher and student triads, research is needed on the overall dynamic of curriculum change.

It is as difficult to describe motion in the model as it is to describe the movement of traffic in a city. Focusing on a single vehicle distracts from the more global view that one gets by looking down on the city and seeing the interaction of all of the vehicles. Similarly, in using the model for organizing research, we must avoid the intensely local views that prevent us from seeing

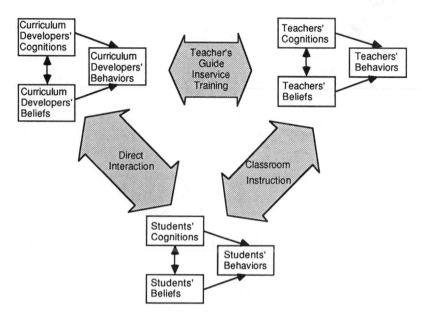

Figure 2. A research model for curriculum development.

the curriculum as a whole. We begin with a knowledge of the "basic traffic flow," prepare the participants for the change, and study the effects of the change as it is adopted. A traffic change made on one street can affect the flow of traffic throughout the city—an improvement made in one aspect of a curriculum may have impact throughout the implementation of the curriculum.

The first step is for the curriculum developers to have a firm theoretical and philosophical foundation for the new curriculum. They should have a clear set of objectives and be able to distinguish how the new curriculum will differ from the existing one. They should also know the degree to which the new curriculum is consistent with the cognitions and beliefs of the students and teachers. Acquiring this knowledge may be the goal of the initial research studies that support and provide direction for a particular curriculum change. One example of such a study in the area of algebra is that conducted by Wagner, Rachlin, and Jensen (1984).

If we assume that the curriculum developers and the teachers who partici-pate in the development process are not the same people, then the developers must share with the teachers some sense of the goals and objectives they hope to accomplish. This communication and interaction should begin early in the project and serve as a foundation for designing the in-service training and teacher's guides to be developed later. Armed with bits of newly formed and unclear theories, fragments of tasks and activities to use in instruction, and their own established cognitions and beliefs, the teachers set out to implement the emerging curriculum with their students.

While some may argue that it is better to study curriculum development in tightly controlled settings, each control adds its own constraint to the eventual application of the curriculum. For example, if one objective of a project is to have the curriculum used in current classroom settings, then even in the initial stages, it is better to develop the curriculum in such a setting. That is, if you expect the curriculum to be used with classes of 30 or more students, then your initial trials should have classes of 30 or more students. It is true that some compromises will no doubt be necessary, but any compromise should be acceded to grudgingly. The classroom interaction between teachers and students provides the next source of research data. These first efforts at implementing the new curriculum in the classroom may provide valuable feedback for adapting and changing the information given to other teachers at later stages of dissemination.

One vital link in the model is early and regular *interaction* between the students and the curriculum developers. Since it is the developers' cognitions and beliefs that drive the change effort, there must be a direct interface between the developers and the students. Whereas traditional field testing tends to be large-scale and post hoc, the recommended interaction may take the form of a teaching experiment in which modifications in instruction and curriculum design are tested at the time of development from direct interaction between the curriculum developer and the students. The modifications suggested by these interactions are then fed back into the development model through the teachers to the students.

One example of the use of research to design a new curriculum is the Hawaii Algebra Learning Project (Rachlin, 1987). Two Algebra I classes were taught in each of the first two years of this NSF-funded project. One class shadowed the other in presentation of content by two days. Each day a student from the first class was interviewed for 45 minutes after class. The interview sample included four students of above-average ability, four students of average ability, and two students of below-average ability (as identified by the classroom teacher). The interviews were videotaped using a split-screen technique that showed both the student and his/her written work. Before class the next day the curriculum developers and the classroom teachers met to view the tape and discuss modifications that might be made in the lesson content and the method of presentation.

As the development process continues, the cognitions and beliefs of the curriculum developers and the teachers regarding the specific goals and objectives of the project change and become more similar. In order to prepare for the dissemination of the curriculum beyond its experimental setting, it is necessary to bring new teachers into contact with the curriculum. Additional research is needed to maximize the success of this teacher preparation, whether the teacher training is through teacher's guides, in-service training, or a combination of the two. After all, the teacher training component is itself a form of curriculum development.

Any model for curriculum research should study change, encourage change, and anticipate change. Each partner in a curriculum triad participates in other triads. For example, a curriculum developer who participates in a wedding of minds like the RAP algebra conference brings new cognitions and beliefs to the curriculum development and dissemination environment. As we learn, our beliefs and cognitions change. The important thing to remember is that there is a research alternative to the traditional curriculum design model and that this alternative should be reflected in the algebra research agenda.

REFERENCES

Bruner, J. S. (1977). *The process of education*. Cambridge, MA: Harvard University Press. (Original work published 1960)

Fennema, E., Carpenter, T. P., & Peterson, P. (1986). *Teachers' decision making and cognitively guided instruction: A new paradigm for curriculum development*. Paper presented at the annual meeting of the International Group for the Psychology of Mathematics Education, London.

Fullan, M. G. (1987). *Implementing educational change: What we know*. Paper presented at the World Bank Seminar, "Planning for the Implementation of Educational Change." (Available from Ontario Institute for Studies in Education, Toronto, Canada)

Osborne, A. R., & Crosswhite, F. J. (1970). Forces and issues related to curriculum and instruction, 7-12. In P. S. Jones (Ed.), *A history of mathematics education in the United States and Canada* (Thirty-second Yearbook, pp. 155-196). Washington, DC: National Council of Teachers of Mathematics.

Rachlin, S. L. (1982). *The processes used by college students in understanding basic algebra*. Columbus, OH: ERIC Clearinghouse for Science, Mathematics, and Environmental Education (SE 036 097).

Rachlin, S. L. (1987). Using research to design a problem-solving approach for teaching algebra. In Sit-Tui Ong, (Ed.), *Proceedings of the Fourth Southeast Asian Conference on Mathematical Education (ICMI-SEAMS)* (pp. 156-161). Singapore: Singapore Institute of Education.

Siemon, D. (1987). Effective strategies for changing mathematics education. *Vinculum, 24*(4), 13-28.

Wagner, S., Rachlin, S. L., & Jensen, R. J. (1984). *Algebra Learning Project: Final report*. Athens: University of Georgia, Department of Mathematics Education.

Part III

Theoretical Considerations

Research Studies in How Humans Think About Algebra

Robert B. Davis
Curriculum Laboratory
University of Illinois at Urbana-Champaign

THIS REALLY IS A NEW WORLD

Something new is happening to mathematics education. That David Wheeler and Jill Larkin and John Thorpe and George Springer and Joan Leitzel and Lesley Booth and Sharon Senk and all of the rest of us should get together in Athens, Georgia, for these sessions is concrete evidence of it. The distances some of us traveled to get here (from Townsville, Australia, for instance) is quite striking evidence. That Herbert Simon, a Nobel prize winner for his work in economics, should spend time studying the teaching and learning of ninth-grade algebra is a provocative piece of evidence. Indeed, Simon has suggested that algebra may do for cognitive science what drosophila did for genetics—provide a convenient problem area of appropriate relevance and complexity, within which important advances may be possible (remarks at the Conference on Cognitive Processes in Algebra, University of Pittsburgh, July 1979).

I offer one more piece of evidence. Working from notes and recollections of the Georgia sessions, I am writing this paper at an indeterminate hour in the early morning, aboard a Boeing 747, en route to Singapore for the Fourth Southeast Asian Conference on Mathematics Education.

Why are busy people taking time to travel to the ends of the earth—almost literally—to conduct serious discussions about how best to study, and improve, the teaching and learning of mathematics?

I suppose the Boeing 747 itself is part of the answer. Take time someday to admire this piece of machinery and to think what was required to produce it. Young people today are growing up in a different kind of world. Even the destination of this flight, on the Asian side of the Pacific, is a clue. This meeting is not being held in Gottingen, though in a earlier age it might have been. (Howson, 1985, has something to say about this.)

Returning—in thought—to Athens, Georgia, and to those very valuable sessions at the algebra conference, I am struck by how many important matters were considered, ranging from "large" general points, all the way up (or down) to highly specific questions, such as what one thinks while solving a linear equation. I want to present some kind of summary of these discussions but, even more important, I want to summarize where we stand as we work on the question of how young people ought to be experiencing— and learning—the subject we call *algebra.*

As I have reworked notes taken at the conference, it seems that what needs to be said can be arranged under three headings: (a) How will students encounter algebra in school? (b) What is mathematics? and (c) How do human beings think about mathematics?

HOW WILL STUDENTS ENCOUNTER ALGEBRA IN SCHOOL?

In the United States, we have fallen into the habit of thinking that "introductory algebra" is a one-year course, usually offered in ninth grade. We also expect that we know the content of this course and the kinds of experiences that will be included. In fact, all three assumptions need to be challenged.

Timing of the Instruction

It is probably still true that the most common pattern in the United States is for algebra to be presented as a one-year course, taught in ninth grade, for which students will have received relatively little prior preparation in earlier grades. In some schools algebra has become an eighth-grade course, in a few (such as University High School in Urbana, Illinois) it has become a seventh-grade course, and there is a project in Lincoln, Massachusetts, that presents algebra in sixth grade. For academically weaker students, some schools slow the pace and offer the usual ninth-grade content in two years, during Grades 9 and 10. Beverly Whittington and her colleagues at Educational Testing Service are developing materials to provide a more effective bridge between arithmetic and algebra, to be presented in Grade 7 or 8. There are those (and I am one) who argue that preparation for algebra should begin in Grade 2 or 3 (Davis, 1985).

Content of the Course

Timing is not the only aspect of algebra where variation occurs—content is another. Some courses combine the usual content of algebra with other topics, including computer programming, geometry, logic, or probability and statistics. In some cases these additional topics are dealt with only briefly, but sometimes they are a major focus of time and attention.

Kinds of Learning Experiences

By far the most important variation in algebra courses occurs in the kinds of experiences that are provided for the students. At one extreme we have the most familiar type of course, where the student is asked to master rituals for manipulating symbols written on paper. The topics in such a course have names like "removing parentheses," "changing signs," "collecting like terms," "simplifying," and so on. It should be immediately clear that a course of this type, focusing mainly on meaningless notation, would be entirely inappropriate for elementary school children; many of us would argue that this type of course, although exceedingly common, is in fact inappropriate for all students.

At what may be the opposite extreme, there is a course based on the kinds of experiences that people like Warwick Sawyer, David Page, and Seymour Papert have often included in demonstration lessons. In this kind of mathematics class, the student starts with some kind of problem or task or goal. Along the way, appropriate notations are created. Most concepts and notations will be the work of students but, even when they are suggested by teachers, their purpose is clear to most students because the goal was present before the notation was introduced. Students and teacher, working together, do what needs to be done.

There seems to be no textbook that follows this approach, but Stein and Crabill (1972) probably comes the closest. Unfortunately, this valuable book was allowed to go out of print. I have written two books attempting this approach (Davis, 1967, 1980), but neither has been much used. More recently, what is essentially this same approach has gained some acceptance when used with computers, as in the style of Logo (see, e.g., Hoyles, 1985). Despite the lack of textbooks, a considerable number of teachers have managed to present courses of this type—and perhaps this kind of course is not one that can easily be based on any textbook. Some suggestion of the sorts of things that creative teachers are doing can be inferred from articles that appear from time to time in the pages of the Mathematics Teacher (see, e.g., Cohen, 1987; Dungle, 1982; Kennedy, 1986; and Wood, 1987). There is more than a little evidence of creative approaches in the recent growth of companies that sell non-textbook supplementary materials (e.g., Camelot Publishing Co., Delta Education, Dale Seymour, Creative Publications, Cuisenaire Co. of America, S. P. Richards, and so on; see also McLoughlin, 1987).

If we want to describe the essential difference between these two types of courses, we might say that in the traditional course "knowledge" is broken up into a large number of small rituals, no larger goals are explicit, the students are shown how to carry out the rituals and are expected to perform them in the orthodox way, the focus is mainly on notation, and essentially no strategic planning or creativity is called for. The meanings of the symbols and of the operations are usually largely ignored.

By contrast, in the Sawyer-Page-Hoyles kind of course, one starts with the goals. It is then a task for the students to devise appropriate strategies for achieving these goals, with key concepts and customary notations arising out of this effort, almost as a by-product.

Even though one may not see a large amount of variation among instructional programs in the United States, those few divergent programs that do exist may hold the best guide to the future. One might attempt to classify alternative algebra courses by how they select among the following possibilities:

- Does the course focus on isolated skills, or does it focus on problems for which the students are asked to devise methods of solution?
- Does the course focus much attention on key concepts?
- Does the course focus much attention on explicit heuristics?
- Are new ideas introduced first via informal or exploratory experiences?
- Do students work together in small groups?
- If students do work in small groups, to what extent is the group responsible for insuring that all members of the group succeed?
- Are algebraic ideas introduced informally in elementary school grades?
- To what extent are multiple representations used, so that the same problem may be seen from algebraic, arithmetical, and geometric perspectives—and perhaps also in concrete or iconic form?

How is the nature of the algebra program relevant for a research agenda? In two ways: First, research cannot take place in a vacuum. Will we study the algebraic ideas of fourth graders, or seventh graders, or ninth graders, or all of these? What content will be taken as the goal? What kinds of skills and understandings will be desired and at what levels of performance? In short, our (possibly implicit) background assumptions about the instructional program will become assumptions for our research program. Second, and conversely, research will help us shape future instructional programs, possibly into quite different forms.

WHAT IS MATHEMATICS?

The matter at stake here is really very important and quite fundamental. It is no exaggeration to say that the central question is "What is *mathematics*, really?" Is mathematics just sets of prescribed directions—what I have deliberately referred to as rituals? There is no question that many people see mathematics as exactly that. If we add to this the fact that, for most students, the rituals seem to serve no sensible purpose and have no recognizable rational basis, it is no mystery that most students find mathematics boring and incomprehensible. Many educators have viewed their task, then, as one of somehow motivating students, somehow "sugar-coating" the subject so as to make it more palatable.

But the nature of mathematics is not the problem at all. These students are not finding *mathematics* distasteful—they are finding these dull, repetitious, essentially *non*-mathematical tasks distasteful. In this, the students are not mistaken. These tasks are not interesting, but they also are not mathematics.

Mathematics is a process of clever analysis of important problems. In doing mathematics one typically starts with a problem—very often a problem that one may not understand very well. By thinking seriously about the problem, one may be able to arrive at a more precise understanding of just what the problem really is. This more precise conceptualization may suggest some possible ways of attacking the problem. In the course of working the problem out, one has to solve many subsidiary problems. For example, one may need to invent a better notation for keeping track of what is going on. One pulls together many techniques and ideas previously used elsewhere.

But this is *not* what a typical ninth-grade algebra course is usually all about. One might attempt to defend the typical course by saying that, before a student can do anything interesting in mathematics, the student must first master the language of mathematics. This is far from the truth and clearly not the way we learn our native tongue—we have many interesting (or even urgent) communication problems with others, and we find effective ways of solving them. I have never heard anyone propose that a child should be taught language by being asked to repeat certain meaningless exercises that are unrelated to the child's real goals. Why, then does anyone imagine that this is the best way to approach mathematics? Yet people do imagine precisely this! Apparently, mathematics has been so badly taught for so many years that we have come to view the traditional approach as inevitable, perhaps even optimal.

An example may make the point more clearly. Some teachers want a student to begin solving the equation

$$2x + 3 = x + 8$$

by thinking, "I'll move the 3 across the equals sign and change its sign." These teachers hope the student will then write:

$$2x = x + 8 - 3.$$

But this approach is surely a case of regarding mathematics as a collection of small, meaningless rituals! Why "move the 3 across the equals sign?" And why on earth should such an act "change [the] sign" of the 3?

Certainly, if we want the student to think of mathematics as consisting of reasonable responses to reasonable challenges, it will be far better if we encourage the student to think, "I can subtract 3 from each side of that equation, without changing its truth set." This may seem like an insignificant change of phrasing, but it is more than that. If the student has seen pictures of balance scales or has worked with actual balances, there can be a very straightforward imagery underlying the idea of subtracting the same thing from each side of an equation.

This leads us to our final main point: How students think about mathematics can be far more important than what, at the moment, they are disposed to write. We must concern ourselves with how students are *thinking* about the subject.

HOW DO HUMANS THINK ABOUT MATHEMATICS?

This question needs to be dealt with on two levels: First, there is the matter of specific instances, similar to the "subtract 3 from each side of the equation" problem that we have just considered. This is undoubtedly where the most progress in mathematics education can be made in the next few years.

But there is a deeper (or more abstract) level. Sooner or later we are going to need a postulated fundamental theory, a basic conceptualization of what people do when they think about mathematics. Clearly, we sometimes store something in memory. Sometimes we retrieve something from memory. Is this all of the basic theory we need—that people can "write into memory" and "retrieve from memory?" Surely not! For one immediate example, we undoubtedly need to know more about those "things" that move in and out of memory. What sort of things are they?

We cannot presently answer these questions, but we need to start thinking about them nonetheless. The development of a postulated fundamental theory of the processes of mathematical thought is surely a task for the years ahead. It is not something already finished that we need only report in these sessions (cf., Davis, 1984).

Mental Representations

What almost certainly must play a major role is the matter of *mental representations*. To think about anything, one must have some sort of mental representation on which to do whatever processing is necessary. That students sometimes do not bother to build much of a mental representation at all has been made clear, for example, by Whitney (1985), as when students report that, if a class of 26 children go on a field trip and the cars in which they travel can carry 4 children per car, then the class will require 65 (!) automobiles for the trip. If those students had made any reasonable mental representation of the problem, they surely would not have envisioned needing 65 automobiles to transport 26 children.

Assimilation Paradigms

Any theory of mental representations will need to consider the primitive building blocks from which more elaborate representations can be constructed. Part of such a theory now exists, in the study of *assimilation paradigms*, or basic metaphors. To illustrate the very helpful role that these metaphors can play in developing mathematics lessons, let us consider an example that one might meet in certain algebra courses. Suppose that one goal of the course is for students to see the possibility of making explicit, careful proofs or derivations, such as the following:

Theorem: $a + a = 2 \times a$

Proof: $2 \times a = 2 \times a$ Reflexive property of equality
         $(1 + 1) \times a = 2 \times a$ Definition of 2
          $a \times (1 + 1) = 2 \times a$ Commutative law of multiplication
  $(a \times 1) + (a \times 1) = 2 \times a$ Distributive law
          $a + a = 2 \times a$ Identity property of 1

From the student's point of view there are several aspects of constructing this proof that can be problematic. Obviously, there is the very interesting question, at each step in the proof, of deciding which step to take next. Of all the things that one could do, which choice will prove useful? In many ways, decision-making is the focus of any cognitive analysis, and quite properly so. The student needs to be well aware of dealing with these choices, and the kinds of cues or clues that can provide guidance are quite interesting from a cognitive science perspective.

There are, however, at least two other aspects of proof construction that often receive less attention. One is the important question of *why* we undertake this kind of task. Too often we fail to deal with such questions. Ask students who have completed high school Euclidean geometry to discuss the historical value of the Greeks' commitment to thoroughly explicit statements and inferences, and you will nearly always find that they have never thought about this aspect of Euclid's approach to geometry. The same sort of thing can happen with the above example in algebra. A good teacher needs to be careful that students understand why one might make a proof of this type.

A second aspect that is problematic but often neglected is the *novelty* of the task. To construct a proof, the student is given a very specific starting point, a very specific end point, and a very specific collection of allowable "actions" or transition rules. Using these rules, the student is asked to get from the starting point to the goal. Has the student ever seen anything like this before? Possibly not. Real life, after all, is hardly ever so specific, so narrowly defined. One could say that chess is somewhat like this, but even in chess the goal is not so fully specified—any old checkmate will do, anywhere on the board, with any piece that can get the job done.

If one believes in assimilation paradigms—or frames, or metaphors (see Cobb, 1987, and Davis, 1984, for a full list of names for this important general concept)—that is to say, if one believes the student understands a new task by retrieving something from memory and mapping this new input data into the retrieved pattern/format/idea/knowledge representation structure, then it is best to make sure the student has something in his or her mind to retrieve. A good mathematics course will put it there.

If I were teaching proofs, I would first have the students work out at least one of those familiar newspaper word puzzles, like

S O U P

— — — —

. . .

— — — —

N U T S,

where you start with the word SOUP, end with the word NUTS, and at each intermediate step have a legitimate 4-letter word. To get from one word in the chain to the next, you may change only one letter.

This is a task most students enjoy. The reason I would use it here is that experiences of this type can help to create an assimilation paradigm for the task of making proofs. The starting point is explicit, the end point is explicit, and the transition rules are explicit. Now it is up to us to find a solution!

Constructing proofs is not the only example of a novel task in mathematics. There are countless situations like this where the nature of what we are doing is not clear to most students. It is important that the teacher recognize this difficulty and try to deal with the problem directly.

I have couched this discussion mostly in the rhetoric of teaching, not the rhetoric of research, but I hope that the research implications are reasonably clear: As we analyze either what the student thinks or what the teacher has done to aid the student's thinking, will we look for the presence of assimilation paradigms of the "what is this task all about" type represented by the soup-to-nuts puzzle? We can look for these paradigms if we choose to do so. Obviously, I would urge that we do, for otherwise we will not know to what we owe student successes and student failures.

WHERE DOES ALL OF THIS LEAVE US?

If we try to put these ideas together, what sort of guidance do they give us for future work? I think that the following assumptions will be crucial: (a) the assumptions we make about the *nature* of the algebra course, (b) the assumptions we make about the kind of *experiences* that will be provided for the students, and especially (c) the assumptions we make about how we want students to *think* about mathematical problems.

If we assume only the standard ninth-grade algebra course, I doubt that we can make much progress. But if we look at the work of those who are creating real alternatives, we can find the materials and assumptions out of which something valuable can grow. I would like to extend Herbert Simon's biological metaphor: Developing the right kinds of mathematics courses is quite a bit like genetic research. If the gene pool is too small, if there is too little diversity, you have no room to experiment and build. But if you can work with a larger gene pool, incredible things can be accomplished.

REFERENCES

Cobb, P. (1987). Information-processing psychology and mathematics education—a constructivist perspective. *Journal of Mathematical Behavior, 6,* 3-40.

Cohen, M. P. (1987). Flexibility and algebraic problem solving. *Mathematics Teacher, 80,* 294-295.

Davis, R. B. (1967). *Explorations in mathematics—A text for teachers.* Palo Alto, CA: Addison-Wesley.

Davis, R. B. (1980). *Discovery in mathematics—A text for teachers.* New Rochelle, NY: Cuisenaire Company of America.

Davis, R. B. (1984). *Learning mathematics: The cognitive science approach to mathematics education.* Norwood, NJ: Ablex.

Davis, R. B. (1985). ICME-5 report: Algebraic thinking in the early grades. *Journal of Mathematical Behavior, 4,* 195-208.

Dungle, J. (1982). The twelve days of Christmas and Pascal's triangle. *Mathematics Teacher, 75,* 755-757.

Hoyles, C. (1985). Developing a context for Logo in school mathematics. *Journal of Mathematical Behavior, 4,* 237-256.

Howson, G. (1985). The impact of computers on mathematics education. *Journal of Mathematical Behavior, 4,* 295-303.

Kennedy, J. B. (1986). Discovering patterns for sums of polygonal numbers. *Mathematics Teacher, 79,* 437-448.

McLoughlin, P. (1987). "Seed problems" in the teaching of mathematics. *Journal of Mathematical Behavior, 6,* 283-291.

Stein, S. K., & Crabill, C. D. (1972). *Elementary algebra—A guided inquiry.* Boston, MA: Houghton Mifflin.

Wood, E. F. (1987). Self-checking codes—An application of modular arithmetic. *Mathematics Teacher, 80,* 312-316.

Whitney, H. (1985). Taking responsibility in school mathematics. *Journal of Mathematical Behavior, 4,* 219-235.

Eight Reasons for Explicit Theories in Mathematics Education

Jill H. Larkin
Department of Psychology
Carnegie Mellon University

Despite a group of interesting papers and many fascinating discussions during the conference, I find a fundamental and, I think, damaging lack of concern for scientific theory in the thinking represented here. I will try to clarify what I believe scientific theory is and what it is not, and why I believe more concern for theory could be beneficial to the field of mathematics education.

WHAT A THEORY IS AND IS NOT

An appropriate theory for mathematics education (or for any field) is not rigid, unbending, or prescriptive of what research must or must not be done. Neither is it a unified whole without alternate parts and views. Indeed, most scientific theories, as they develop, take bits and pieces from many sources. A good example is quantum mechanics, originally based on ideas of musical harmonics, electrical interaction, and planetary motion. The nature of theory is change—today's theory is always the best we can do to account for today's data, but major or minor changes will be required if tomorrow's data contradict it. Theories are not even the dominant force in guiding the design of new experiments, especially in a young field like mathematics education. Simon and his colleagues (Langley, Simon, Bradshaw, & Zytkow, 1987) argue that science progresses largely by noticing patterns in data. Finally, it should hardly be necessary to point out to a group so strongly based in constructivism that a theory is not "truth" or "reality." It is a construction existing in the minds of scientists that helps them in their work.

WHY MATHEMATICS EDUCATION NEEDS GOOD THEORIES

I believe that formulating theories to account for data is absolutely essential to scientific progress and that theories are most crucial in difficult areas such as mathematics education. My main arguments are the following:

1. A theory is explicit and, within its own domain, coherent. Struggling to write even partial theories exposes defects and incompleteness in our thinking. Refusing to write theories allows such defects to go unchallenged.

2. Trying to write a theory forces us to attempt to unify as much data as possible under a single framework. This process clarifies relations among data that may otherwise go unnoticed.

3. For good reason, science values theories that are parsimonious or simple. Humans have only very limited reasoning capacities. By tying disparate results together under a simple, clear framework, we help ourselves to establish new relationships among otherwise unconnected results.

4. I believe theories should be written in simple language that can be understood by the widest possible audience. Jargon and technical or specialist terms prevent others from using a theory and, as stated above, theories are often constructed in bits and pieces from other theories. Even worse, obscure language can prevent the clear thinking that theories should promote.

5. If experiments are reported together with their connection to existing theory (either supportive or contradictory), then the scientists who write the reports are telling readers clearly what they think the experiments mean and how they relate to other knowledge. Readers need not agree, but at least they are engaged in thoughtful dialogue, rather than being bombarded with more isolated results.

6. Stating theoretical implications is an excellent way to suggest what information others might use. When scientists run an experiment, they do other scientists a great service by communicating their best judgment as to what parts of the results are general and might be applicable elsewhere, and what parts are not. Again, the readers need not agree but, as the persons most knowledgeable about the experiment, scientists should make their judgments available.

7. When one scientist suggests that a result may be general, and others confirm or contradict the result in other contexts, theories build and become increasingly useful, only because we have growing confidence that the knowledge can be used in many situations. (Of course, no theory is guaranteed to be applicable in any future research.)

8. Perhaps most importantly, formulating theories that clearly and succinctly relate data is challenging, creative work. It makes us think hard about what we know and pushes our curiosity to inquire further about what we do not know.

For all of these reasons, I feel a certain disappointment in much of the work collected in this volume. I believe hard creative thought would make the theories more interesting, more encompassing, and more suggestive of useful new work. I am saddened by the wide use of special terminology which may prevent outsiders from (a) understanding the results of mathematics education research, (b) using these results as inspiration for their own work, and (c) participating in discussions with mathematics education researchers.

But mathematics education is a young field, and the study of algebra learning is even younger. It is full of fascinating questions and lively investigators, and perhaps a few more years of exploration will encourage more of the disciplined but creative thinking for which I have tried to argue.

REFERENCE

Langley, P., Simon, H. A., Bradshaw, G. L., & Zytkow, J. M. (1987). *Scientific discovery: Computational explorations of the creative processes*. Cambridge, MA: MIT Press.

Contexts for Research on the Teaching and Learning of Algebra

David Wheeler
Department of Mathematics
Concordia University, Montreal

THE GULF BETWEEN RESEARCH AND PRACTICE

One of the assumptions of the Research Agenda Project is that research on mathematics education can make a difference to mathematics instruction. It is not obvious that this assumption is justified, as I have written elsewhere:

> The separation of researchers from practitioners, and of theory from practice, is common to most professional activity in many societies. The structure of academic institutions, which distinguish "pure" and "applied" science, and institutionalized differential career opportunities, reinforce the separation. In education the gap is particularly wide and distressing. This is not so much because teachers do not want technical help, nor because researchers would not be able to supply it, but because each group is embedded in a different situation with its own goals, responsibilities, and rewards, none of which relate to communication between the two groups. Researchers are not recognized by the extent to which their work proves useful to teachers, nor are teachers recognized by the extent to which they are informed about and use the latest research. The system of separation, meant to be rational and efficient, does not work. (Wheeler, 1986, p. 57)

This unfortunate situation, which is discussed in its application to other professional contexts by Schön (1983), must be taken into account as one of the factors necessarily influencing the impact of the Research Agenda Project. Even if the Agenda were to be endorsed and acted upon by the community of mathematics education researchers, it is not clear to what extent mathematics instruction would be touched, if at all.

THE FIELD OF MATHEMATICS EDUCATION

Some attempts to counter the separation between research and practice seem to be taking place. The last few years have seen a noticeable increase in the number of references to "mathematics education" as a field of study and development which is of professional interest to teachers of mathematics, teacher educators, mathematicians, educational researchers, and administrators. The convergence of these groups is symbolized and to some extent facilitated by, for example, the *Journal for Research in Mathematics Education*, published by the National Council of Teachers of Mathematics— though occasional controversies about the role and style of the journal show that the task of bridging professional gaps is still problematic.

Twenty years ago the quadrennial International Congresses on Mathematical Education were launched; springing out of them have come a number of ongoing groups, including the International Group for the Psychology of Mathematics Education. Twenty-five Instituts de Recherche sur l'Enseignement des Mathématiques (IREMs) have been formed in France. The Federal Republic of Germany has an Institut für Didaktik der Mathematik and the United Kingdom, two Shell Centres for Mathematical Education. There are several concentrations of researchers in mathematics education in the United States and similar developments elsewhere, including a vigorous centre in Papua New Guinea. All this activity does not necessarily signal a "brave new world," but it does indicate the relatively recent determination of some educators to establish mathematics education as a field of research and development in its own right.

Mathematics education borrows concepts, theories, knowledge, and research techniques from psychology, sociology, linguistics, cognitive science, and so forth, but the questions it works on tend to be strongly connected to the subject matter of mathematics. This is not such a trite observation as it may seem. Earlier research into matters of interest to teachers of mathematics was much more likely to be seen as part of general educational psychology, involving constructs about human abilities, for instance, or generalized learning theories. Such researches tended to take mathematics as an uncontroversial "given."

The reasons for such a decided change in research emphasis as has taken place are much more complex than can be discussed here, but one can certainly point to the influence of Piaget, who delineated the intrinsic difficulties in the acquisition of some scientific concepts. The development of powerful computers made the goal of simulating human intellectual behaviour seem realizable, and the detailed investigations of human thinking and problem solving which followed showed that this behaviour is highly task-oriented. Both approaches have generated many research questions centring on the interaction between human beings and mathematical tasks. In principle, this change in research direction would seem to be potentially very helpful to curriculum planners, textbook writers, and teachers at all levels of instruction.

REMARKS ON THE HISTORY OF ALGEBRA

Historically, the development of algebra as a symbol system took a very long time indeed. Without attempting to describe this evolution in detail, it is sufficient to make two points about the language of high school algebra that are particularly important: (a) The formal symbolism is not essential for handling some of the problems most closely associated with it, and (b) the development of a specialized symbolic language is a result of an act of abstraction in which the meanings of individual items, and even of the operations acting on them, may be almost completely eliminated. Symbolic algebra is semantically very weak (Wheeler & Lee, 1986).

The debate among British mathematicians in the first half of the 19th century about the nature of algebra draws attention to another important aspect of algebra—an epistemological problem. One side took the position that algebra is "universal arithmetic"—that is, it deals with quantities and permissible operations on quantities, and its rules are dictated by the well-known properties of quantitative arithmetic (Pycior, 1984). The position of those on the other side of the argument was that algebra is a purely symbolic system; it deals with essentially arbitrary symbols governed by essentially arbitrary rules. The extreme positions on either side of the argument may now seem untenable to us, but enough heat was generated at the time to suggest that a more difficult question than the one apparently being discussed may have been embedded in the situation: If the rules of algebra do not necessarily derive from the behaviour of numbers and quantities, where do they derive from? Saying that they are arbitrary carries little conviction, and mathematicians do not behave as if this were the case, so this formulation seems to be merely a way of asserting that algebra can be associated with the behaviour of things other than measurable quantities.

ALGEBRA IN THE SCHOOL CURRICULUM

The fact that we know that modern algebra is not irrevocably tied to arithmetic does not by itself imply that we must change the way students are introduced to algebra. For them, it may be best that they start from the familiar behaviour of the numbers that they know well. But history reminds us that algebra is very much more than generalized arithmetic, and we should be more realistic in our approach to algebra. We should not assume that the transition from arithmetic to algebra is obvious and clear sailing.

> The time has come for a careful reappraisal of the aims and content of algebra courses and of ways of teaching the subject. In any case the teaching of traditional algebra has long presented difficulties in schools and it is a branch of mathematics which remains a mystery to many adults. (Her Majesty's Inspectorate, 1979)

These remarks, from a report of the Government education inspectorate in England and Wales, could be echoed in most countries. Algebra has a firmly established place in the mathematics curriculum for most students in most countries but, in spite of the many decades during which it has been taught, the pedagogical problems it presents have by no means been solved. The elementary stages of a high school algebra course are the end of the mathematical road for some students; the struggle to learn algebra gives them a negative message about their ability to learn mathematics of any sort. Even those who survive the beginnings of algebra do not generally understand until later what algebra is able to do, what it is "for." Indeed, algebra is traditionally taught for its importance as a tool needed to handle the mathematics that is to come later—trigonometry and calculus, for instance—rather than as a branch of mathematics with a use and character of its own. Putting its justification into the future in this way forces students to approach it as something whose purpose and relevance to their own interests is not yet clear. While this demand made by schools is by no means restricted to algebra, the abstract nature of algebra and the unusual form of its symbolism make the necessary suspension of judgment about the value of the activity very difficult for learners to achieve.

The adoption of algebra into the curriculum took place, first at the undergraduate level and later at the high school level, without the pedagogical consequences of the "universal arithmetic" versus "symbol system" dichotomy being resolved: To what extent is beginning algebra an abstraction from and a generalization of quantitative arithmetic? To what extent is it a distinct symbol system whose rules can be determined by convenience? The traditional teaching of beginning algebra contains elements of both approaches and, as a consequence, novice learners of algebra are often confused about the proper source of validation of what they do. Is a certain transformation correct because it corresponds to the ways that numbers and number operations behave, or is it correct because it obeys the appropriate "arbitrary" rule? Perhaps we should not be surprised at students' failure to grasp in one year what it took mathematicians over 3500 years to clarify.

Freudenthal (1983) points out that the traditional teaching of algebra places on students the demand to become fluent in handling the formal algebraic language, while at the same time understanding how this language may be used to solve meaningful problems. The two aspects of this demand are difficult to meet simultaneously, and it is not surprising that teachers and textbooks are better at the easier task—manipulating the formal language. There may not be a vast number of people around who can, say, formally solve a quadratic equation, but there are vastly more who can do this than who know when and how to apply the technique to the solution of some genuine problem. Although, in a global sense, one can say that algebra plays a part in the solution of enormously many real problems—algebra is needed, as someone at the conference said, to make it possible to mathematize the world—it is still very difficult to find problems that can be used to show high school students how and when to apply the algebra that they know (Fey, 1984). An inspection of school algebra textbooks shows that most of the problems they contain, whether "real" or artificial, would be better handled, and certainly better understood, if treated as problems to be solved by arithmetic.

A fresh look at the contents, place, and purpose of high school algebra seems long overdue. Americans might take note of experiences in other countries where algebra is handled differently in the curriculum. They might also be especially cautious about the advice they get from mathematicians. During this century, on the two occasions when mathematicians' voices have been particularly attended to—namely, at the beginning of the century, when algebra courses were being established, and again in the 1960s, when the "new math" wave was passing through—their advice pushed schools further in the direction of formalism. One sees why. In line with their own self-interest, they saw that undergraduate students needed to be fluent in algebraic techniques and rigorous in their use of them. The mathematicians did not see, perhaps because it had not happened to them, that this highly technical preparation might kill the mathematical confidence of a lot of able students. Or, if they did see it, they approved of it as a sort of "survival of the fittest" process.

Could we not develop algebra courses that meet more complex goals than just giving potential undergraduates the technical competence their professors would find it convenient for them to have?

> Algebra is now not merely "giving meaning to the symbols," but another level beyond that: concerning itself with those modes of thought that are essentially algebraic—for example, handling the as-yet-unknown, inverting and reversing operations, seeing the general in the particular. Becoming aware of these processes and in control of them, is *what it means to think algebraically* [italics added]. (Love, 1986, p. 49)

This may sound flabby in comparison with the well-defined, technique-laden algebra curriculum we know now, but by thinking along these lines we might be able to produce courses that would have more appeal and relevance to the 60% of students who are almost totally baffled by what they are offered now. Of course, as Nel Noddings (1986) reminds us, it is unlikely that the politics of curriculum decision-making in America would favour, or even permit, such a radical change in direction. She suggests that we could use the idea of such a change, though, as a heuristic device to help us consider how to ameliorate the present curriculum.

In concluding this section, it may be worth making the general point that what is good for mathematics is not necessarily good for pedagogy. The "new math" is one case in point; a much earlier example is provided by the advent into Europe of the Hindu-Arabic numeration system. When this system was generally adopted and entered the schools, the former concrete methods of introducing counting and calculating through the use of beads and other objects were virtually abandoned and replaced by instruction in "ciphering"—which the very excellence of the place-value numeration system permitted. In this way, meaning was stripped from the activities of learning to count, add, multiply, and so forth (Smith, 1900).

SOME REFLECTIONS ON THE CONFERENCE DISCUSSIONS

Nothing very radical was said about algebra as an item in the school curriculum, or about algebra as an item in the mathematician's repertoire. The latter omission could have been expected: People came to the conference to discuss algebra in education, not to sort out and make sense of the place of algebra in the edifice of mathematics. But an opportunity to scrutinise algebra itself may have been missed. Dewey said somewhere that subject matter is a prime source of pedagogical insights. Almost no educators really believe this, I think, except in the trivial sense of hoping that teachers, textbook writers, and curriculum designers "know their mathematics." Even many mathematicians, who ought to know better, have no interest in looking below the instrumental or formal surface of mathematics in order to get clues about how to present it more effectively. That mathematicians *can* look at mathematics in such a way as to generate pedagogical insights is attested to by the examples of Dienes, Freudenthal, Gattegno, and Polya, but the few who do are extremely rare birds.

In the curriculum, algebra still seems enthroned in its peculiarly North American position of dominance. Nobody at the conference seriously suggested, for example, that it might be taught earlier, to fewer students, integrated with geometry, or any of the other options practised in other countries. It seems to me that many of the problems which appear to be problems about the learning and teaching of algebra are, in fact, relatively local: They are problems about learning and teaching algebra in the American school system.

The central importance of the function concept to algebra courses in school is one of those assertions that needs to be looked at carefully. We all know how important the function concept is—and the magnitude of the breakthrough that resulted when the idea was first entertained—so it is hard to refrain from nodding vigorously when someone says that this is what school algebra should be about. But the function concept, as well as being important, is also subtle and complex, as shown by the 200 years mathematicians took trying to define it. With hindsight it is easy to see that a lot of "pre-function" mathematics can be recast in terms of functions, but the crucial pedagogical question is whether beginning students are better off working at this level of abstraction and generality right from the start or whether they should be allowed to approach it gradually. Certainly, if curriculum designers and textbook writers are told that Algebra I and II are really all about functions, functions in all their formal glory will be there in the next generation of curricula and textbooks, and almost everything else will have disappeared. (By the way, when *does* the American high school student get the chance to investigate some mathematical situations, solve some genuine mathematical problems, begin learning how to make conjectures, and put arguments together in the form of a proof?)

In principle, the fact that calculators and computers are now available to take the drudgery out of doing algebra ought to make students, teachers, and everyone involved, shout for joy. Members of the conference seemed to feel that shouting was premature and that there were questions to answer first. In addition to the obvious questions, such as:

- What is it now unnecessary to teach?
- What can now be taught that could not be taught before?
- By what new approach can we now teach such-and-such?

there are less obvious ones, like:

- To what extent is algorithmic performance related to understanding?

The prototype curriculum project directed by James Fey (which contains many innovative and exciting elements) starts with "zero-based" planning decisions about the content of the curriculum. Everything in the current curriculum is assumed to be out and must win a place on its own merits. In practice, as one might expect, "merit" seems to be synonymous with "utility." Even with a generous interpretation of the scope of this criterion, instrumentality prevails. A curriculum item will remain banished if its only claim to consideration is, say, aesthetic or recreational or cultural.

Utility! Instrumentality!! Eventually, it seems, every discussion about the practice of education—teaching, pedagogy, curriculum—gets pulled down into this apparently irresistible vortex where nothing survives unless it can demonstrate that it is an essential prerequisite for doing something else (especially in North America, though not only here). All the suggestions about the algebra curriculum made at the conference (except perhaps those by David Tall) seemed to me to remain within the same old frame of reference: Any proposal has to justify itself by its projected usefulness.

It should be possible to take a more flexible attitude and have more subtle discussions without slipping into total unreality, to allow non-utilitarian valuations while still accepting that utility is a most important consideration. It would mean being more careful not to lump together, as if they were equivalent, the valuations that may be put on mathematical activities by the receivers of instruction, the givers of instruction, and the providers of instruction (viz., society) and not to suppose that the individual valuations within each group are all the same. As Otte (1985) says, "From the point of view of society as a whole, all knowledge has an instrumental and not an autotelic value" (p. 5). Knowledge contributes to the life and health of that society; it is instrumental to its well-being. A healthy society will give considerable autonomy to its citizens to decide their own goals and conditions of life, so the knowledge that is of direct use to some is not to others. Theoretical knowledge, which has instrumental value for only a handful of people, may nevertheless open up a universe of experience rich enough to allow the participation of all members of society (Otte, 1985). Indeed, an education that remains always at the level of "only utility counts" is in a poor position to fulfill any goals, even utilitarian ones. We cannot understand the part played by algebra in permitting the mathematization of the world if we are only shown a few of the "useful tricks" and are never exposed to its more general strategies or given the chance to reflect upon them.

The educator who is also a researcher cannot always keep the two personas in step. Educators have to think about problems and take actions to solve them without waiting for researchers to tell them what to think and what to do. Researchers, on the other hand, are well advised to take as much time as they need (if such anti-educational policies as the annual counting of publications do not force them to behave otherwise). The researcher's problem, in addition to the tactical one of deciding what a granting agency may be willing to support, is that of finding some enquiry to pursue which is both promising and manageable. The latter tends to become the dominant criterion, especially if the researcher is working essentially alone. Many of the researches admirably surveyed by Carolyn Kieran (this volume) and Nicolas Herscovics (this volume) are small-scale in scope, duration, and significance. Is there anywhere among them a single paper which momentarily shocks a reader into a new state of perception of some phenomenon? Well, maybe, and maybe not. There is journeyman research in all fields, large in quantity, and often useful in making the terrain more inhabitable. But where are the researches that actually claim some fresh territory? More of those, please.

FRAMEWORKS FOR RESEARCH

Since mathematics education is not a discipline, with all the attendant benefits of agreement on logic and heuristic, it seems unlikely that there could be a single, comprehensive theory that would provide a framework for research about the learning and teaching of mathematics. Relatively little published research in the area is theory-driven. Perhaps Piagetian research, particularly when designed by the master himself, is strictly governed by a theory. The principles of artificial intelligence ensure that a certain style of research is done in its name. But neither "information processing" nor "constructivism" (which were the two candidates for governor most frequently mentioned at the conference) come equipped with a sufficiently well-defined programme. The eclectic collections of theoretical assumptions associated with each of them do not amount to a framework, as far as I can see.

Some very good research gets done without a heavy freight of theory—and it is worth remembering that theory is a constraint as well as a support. Perhaps it is more realistic to think about potentially soluble problems whose solution would be likely to advance mathematics education. When one asks people to propose such problems (see the three issues of Volume 4 of *For the Learning of Mathematics*), the differences in the suggestions seem to come not from the espousal of different theories but from a focus on different parts of the field and from different choices about the kinds of evidence to be taken into account. Theoretical differences are certainly there, implicitly or explicitly, but it is the differences in heuristic that seem to be more important.

Mathematics education research would be strengthened, I believe, by extending the set of heuristics that researchers could call on to include components related to (a) mathematics and (b) the teaching situation. Educators, by definition, are concerned with people appropriating mathematical skills, concepts, and knowledge: Their students are "coming to know" mathematics. Part of what the students have to do to get to know mathematics is similar to what they have to do to get to know anything else (and these are the processes of particular interest to the cognitive psychologist), but the character of mathematics is quite distinctive, and coming to know it also makes distinctive demands. A "genetic epistemology" of mathematics gives researchers—and teachers, too—information about the demands that particular mathematical tasks place on any learner.

Patrick Thompson (this volume) made the important point in his talk that the errors students make are to some extent a function of the way they are taught and are not exclusively determined by the nature of the tasks they are working on. The kind of "buggy" behaviours that students exhibit when working on algebraic skills may differ depending on the way the tasks are presented to them—either as structured by rules or as abstracted from the behaviour of quantities. This point complements—it does not contradict—the one above. Mathematical tasks do largely determine the behaviour of people grappling with them; but tasks are always tasks-in-some-context, and the nature of the context, or the learner's perception of the context, also becomes a component of the determining matrix.

Mathematics education research should be sensitive to context, and particularly to didactical intention, as a significant contextual element. It is unfortunate that the word *didactics* does not really exist in American educators' vocabularies—as if not using a word might make the thing it refers to disappear. M. Jourdain spoke prose without knowing it, and American educators follow didactical principles without admitting to them. Perhaps this is why educational research in America for the most part ignores the peculiarly effective power of the triadic teacher-learner-subject matter interactions that anyone would suppose must be at the very core of every educational endeavour.

REFERENCES

Fey, J. T. (Ed.). (1984). *Computing and mathematics: The impact on secondary school curricula.* Reston, VA: National Council of Teachers of Mathematics.

Freudenthal, H. (1983). *Didactical phenomenology of mathematical structures.* Dordrecht, The Netherlands: D. Reidel.

Her Majesty's Inspectorate. (1979). *Aspects of secondary education in England: A survey by HM Inspectors of Schools.* London: Her Majesty's Stationery Office.

Love, E. (1986). What *is* algebra? *Mathematics Teaching, 117,* 48-50.

Noddings, N. (1986). The heuristics of teaching. In G. Lappan & R. Even (Eds.), *Proceedings of the Eighth Annual Meeting of the North American Chapter of the International Group for the Psychology of Mathematics Education* (Vol. 2, pp. 43-64). East Lansing: Michigan State University.

Otte, M. (1985). *Mathematics, computers, and schools from an epistemological perspective* (Occasional paper 68). Bielefeld, FRG: Institut für Didaktik der Mathematik der Universität Bielefeld.

Pycior, H. M. (1984). Internalism, externalism and beyond: 19th-century British algebra. *Historia Mathematica, 11*, 424-441.

Schön, D. A. (1983). *The reflective practitioner: How professionals think in action.* New York: Basic Books.

Smith, D. E. (1900). *The teaching of elementary mathematics.* New York: Macmillan.

Wheeler, D. (1986). The teaching of primary and secondary school mathematics. In D. Layton (Ed.), *Innovations in science and technology education.* Paris: UNESCO.

Wheeler, D., & Lee, L. (1986). Towards a psychology of algebra. In G. Lappan & R. Even (Eds.), *Proceedings of the Eighth Annual Meeting of the North American Chapter of the International Group for the Psychology of Mathematics Education* (Vol. 1, pp. 133-138). East Lansing: Michigan State University.

Working Group on
Teaching and Assessing Problem Solving

San Diego, California   January 9–12, 1987

Joan Akers
San Diego County
Office of Education

John Bransford
Vanderbilt University

George W. Bright
University of Houston

Ann L. Brown
University of Illinois

Thomas P. Carpenter
University of Wisconsin-
Madison

**Randall I. Charles
Illinois State University

Clyde Corcoran
Whittier High School
District

John Donald
San Diego State University

*James G. Greeno
University of California,
Berkeley

*Jeremy Kilpatrick
University of Georgia

Gerald Kulm
AAAS

Jean Lave
University of California,
Irvine

Frank K. Lester
Indiana University

Sandra P. Marshall
San Diego State University

*Douglas B. McLeod
Washington State University

Nel Noddings
Stanford University

Nobuhiko Nohda
University of Tsukuba

Tej N. Pandey
California Assessment
Program

Lauren B. Resnick
University of Pittsburgh

*Thomas A. Romberg
University of Wisconsin

Alan H. Schoenfeld
University of California,
Berkeley

Richard J. Shavelson
University of California, Los
Angeles

**Edward A. Silver
San Diego State University

*Judith T. Sowder
San Diego State University

Larry Sowder
San Diego State University

George M. A. Stanic
University of Georgia

*James W. Stigler
University of Chicago

Alba G. Thompson
Illinois State University

James W. Wilson
University of Georgia

Working Group on
Effective Mathematics Teaching

Columbia, Missouri   March 11–14, 1987

Heinrich Bauersfeld
Universitat Bielefeld

Jacques C. Bergeron
Université de Montréal

David C. Berliner
University of Arizona

Bruce J. Biddle
University of Missouri

Catherine A. Brown
Virginia Polytechnic Institute

Stephen Brown
State University of New
York-Buffalo

William S. Bush
University of Kentucky

**Thomas J. Cooney
University of Georgia

John A. Dossey
Illinois State University

Elizabeth Fennema
University of Wisconsin-
Madison

Sherry Gerleman
Eastern Washington State
University

Thomas L. Good
University of Missouri

**Douglas A. Grouws
University of Missouri

Celia Hoyles
University of London

Martin L. Johnson
University of Maryland

Mary Koehler
University of Kansas

Perry E. Lanier
Michigan State University

Gaea Leinhardt
University of Pittsburgh

Richard Lodholz
Parkway Public Schools

Marilyn Nickson
Essex Institute of Higher
Education

John Owens
University of Alabama

Penelope L. Peterson
University of Wisconsin

Andrew C. Porter
Michigan State University

Edward Rathmell
University of Northern Iowa

Laurie Hart Reyes
University of Georgia

*Thomas A. Romberg
University of Wisconsin

Janet W. Schofield
University of Pittsburgh

Robert Slavin
Johns Hopkins University

*Judith T. Sowder
San Diego State University

*James W. Stigler
University of Chicago

Working Group on
The Learning and Teaching of Algebra

Athens, Georgia   *March 25–28, 1987*

John E. Bernard
West Georgia College

Lesley R. Booth
James Cook University-
Queensland

Diane J. Briars
Pittsburgh Board of
Education

Seth Chaiklin
Bank Street College

Robert B. Davis
University of Illinois

James T. Fey
University of Maryland

Eugenio Filloy Yague
Centro de Investigacion y
Estudios Avanzados del
I.P.N.

Larry L. Hatfield
University of Georgia

Nicolas Herscovics
Concordia University-
Montreal

Robert Jensen
Emory University

Mary Grace Kantowski
University of Florida

James J. Kaput
Southeastern Massachusetts
University

**Carolyn Kieran
Université du Québec a
Montréal

David Kirshner
University of British
Columbia

Jill H. Larkin
Carnegie Mellon University

Joan R. Leitzel
Ohio State University

Matthew Lewis
Carnegie Mellon University

Tatsuro Miwa
University of Tsukuba

Sidney Rachlin
University of Hawaii

Sharon L. Senk
Syracuse University

*George Springer
Indiana University

*Judith T. Sowder
San Diego State University

*Jane O. Swafford
Northern Michigan
University

David Tall
University of Warwick

Patrick W. Thompson
Illinois State University

John A. Thorpe
National Science Foundation

**Sigrid Wagner
University of Georgia

David Wheeler
Concordia University-
Montreal

Patricia S. Wilson
University of Georgia

Working Group on
Middle School Number Concepts

DeKalb, Illinois   *May 12–15, 1987*

*Merlyn J. Behr
Northern Illinois University

Alan Bell
Shell Centre, Nottingham

Robbie Case
Ontario Institute for Studies
in Education

Karen C. Fuson
Northwestern University

Brian Greer
Queens University, Belfast

Kathleen M. Hart
Kings College, London

*James Hiebert
University of Delaware

Thomas E. Kieren
University of Alberta

Magdalene Lampert
Michigan State University

Glenda Lappan
Michigan State University

Richard Lesh
WICAT

Jack Lochnead
University of Massachusetts

*Douglas B. McLeod
Washington State University

Pearla Nesher
University of Haifa

Stellan Ohlsson
University of Pittsburgh

Joseph N. Payne
University of Michigan

Thomas R. Post
University of Minnesota

Robert E. Reys
University of Missouri

*Thomas A. Romberg
University of Wisconsin

Judah L. Schwartz
Education Technology
Center

*Judith T. Sowder
San Diego State University

Leslie P. Steffe
University of Georgia

Gérard Vergnaud
Greco Didactique, Paris

Ipke Wachsmuth
Universitat Osnabruck

Diana Wearne
University of Delaware

Advisory Board members and Project Director
Conference Co-directors